Fundamentals of Soil Science

About the Author

Dr. A. Raja Rajan holds a doctorate in Soil Science & Agricultural Chemistry from Tamil Nadu Agricultural University (TNAU), Coimbatore. He had his post-doctoral training at University of Florida, Gainesville, USA as an FAO Fellow. He has served in various positions in TNAU and was Dean at Agricultural College & Research Institute, Killikulam (TN), Dean, Faculty of Agriculture, TNAU, Coimbatore and Dean, Faculty of Agricultural Education and Research of Ramakrishna Mission Vivekananda Educational and Research Institute at its Coimbatore Campus. With vast teaching and research experience in Agriculture, spanning over 40 years, he has guided several post-graduate and doctoral students and has 105 research publications, four books and five technical reports to his credit. He is an expert in the use of nuclear techniques in Agricultural research, with specialized trainings at Bhabha Atomic Research Centre, Trombay, University of Florida and University of Georgia and has handled several research projects involving radiotracers in the disciplines of soil fertility, soil chemistry and environment.

Fundamentals of Soil Science

A Raja Rajan

Formerly
Dean, Faculty of Agriculture
Tamil Nadu Agricultural University
Coimbatore - 641003, Tamil Nadu, India
and
Dean
Agricultural College & Research Institute
Killikulam - 628252
Tamil Nadu, India

NEW INDIA PUBLISHING AGENCY
New Delhi-110 034

nipa

NEW INDIA PUBLISHING AGENCY
101, Vikas Surya Plaza, CU Block, LSC Market
Pitam Pura, New Delhi – 110 034, India
Email: info@nipabooks.com
Web: www.nipabooks.com

For customer assistance, please contact
Phone: + 91-11-27 34 17 17 Fax: + 91-11- 27 34 16 16
E-Mail: feedbacks@nipabooks.com

ISBN: 978-93-90591-44-2

Composed and Designed by NIPA.

Dedicated to

My Professor

Dr. U.S. Sree Ramulu Ph. D. (Calif.)

with gratitude and reverence

Preface

Soil is a finite resource and its loss and degradation is not recoverable within a human lifespan. Soils affect the food we eat, the water we drink, the air we breathe and our overall health and the health of all organisms on the planet. Without healthy soils, we would not be able to grow our food, or meet our fibre and fodder requirements. It is estimated that 95 per cent of our food is directly or indirectly produced on our soils.

Since soil is one of the most complicated biological materials on our planet, it is imperative that we care for soil and nurture it the way it should be for the future generations. In this context, we all must have a basic understanding of the soil, its nature, properties and functions so that we will appreciate it and care for it.

It is with this objective that this book is written. It has been attempted to present in a logical sequence all basic information about soils, from the definition of soil to how they are formed, the factors and processes governing their formation, their intrinsic properties, classification, functions, and the role they play in the global environment today and for a sustainable tomorrow. Emphasis on soil quality and soil health care has been laid as it is the order of the day. The 32 chapters of the book will provide a wealth of information for those looking to take their first deep dive into the realm of soils!

I earnestly hope this book provides in simple, lucid style all necessary information that the students and teachers of Soil Science look for. Suggestions to improve the contents and presentation of the book are most welcome.

I would like to express my sincere appreciation of the wonderful support given by my family during the writing of this book and to New India Publishing Agency for bringing out this book in an impressive way.

Bengaluru, India **A. Raja Rajan**

Contents

1

Introduction to Soil

"SOIL" the term is derived from Latin word "solum" which means, "floor". Soil has been a defining component of humanity since the beginning of civilization. It holds the clues of past cultures, it supports the web of life connecting all ecosystems, provides materials to build houses, and provides the base for our food, feed, fiber and even some biofuels.

Soils are complex mixtures of minerals, water, air, organic matter, and countless organisms that are the decaying remains of once-living things. It forms at the surface of land and is the "skin of the earth." Soil is capable of supporting plant life and is vital to life on earth. There are several general roles that soils play:

- Soils serve as media for growth of all kinds of plants.
- Soils modify the atmosphere by emitting and absorbing gases (carbon dioxide, methane, water vapor, and the like) and dust.
- Soils provide habitat for animals that live in the soil (such as groundhogs and mice) to organisms (such as bacteria and fungi), that account for most of the living things on Earth.
- Soils absorb, hold, release, alter, and purify most of the water in terrestrial systems.
- Soils process recycled nutrients, including carbon, so that living things can use them over and over again.
- Soils serve as engineering media for construction of foundations, roadbeds, dams and buildings, and preserve or destroy artifacts of human endeavors.
- Soils act as a living filter to clean water before it moves into an aquifer.

1.1. Definitions of Soil

Soil is looked upon by mankind in different perspectives. For a layman, soil is a dirt; for an agriculturist, it is a habitat for plants; for a mining engineer, it is debris covering the rocks that is to be removed; for a civil engineer, it is a surface on which road is to be laid; for a child, it is play field. There are different ways in which soil can be defined in a scientific way. Most simple definition would be that *"soil is the natural medium of plant growth"*. It can also be described as a mixture of rock and mineral material with organic matter.

Eugene W. Hilgard (1833-1916), Germany-born American expert on Pedology who is considered in the United States as the Father of modern Soil Science defined soil as *"the more or less loose and friable material in which, by means of their roots, plants may or do find a foothold and nourishment as well as other conditions of growth"*.

Vasily V. Dokuchaev (1846-1903), the Russian Geologist, who is credited with laying the foundation for Soil Science and is called the Father of Soil Science, defined soils as *"independent natural bodies, each with a unique morphology resulting from a unique combination of climate, living matter, earthy parent materials, relief and age of land form"*.

V. V. Dokuchaev

Jacob S. Joffe (1886-1963) defined soil as *"a natural body of mineral and organic constituents differentiated into horizons which differ among themselves as well as from the underlying material in morphology, physical makeup, chemical composition and biological characteristics"*.

Soil is formally defined in the Soil Science Society of America glossary of Soil Science as:

(1) the unconsolidated mineral or organic material on the immediate surface of the earth that serves as a natural medium for the growth of land plants.

(2) The unconsolidated mineral or organic matter on the surface of the earth that has been subjected to and shows effects of genetic and environmental factors of: climate (including water and temperature effects), and macro- and microorganisms, conditioned by relief, acting on parent material over a period of time.

1.2. Approaches to Study of Soil

There are two approaches to the scientific study of soil. One treats soil as a natural body, weathered and synthesized product in nature (Pedological approach). The other approach treats soil as a medium for plant growth and studies soil from the standpoint of higher plants (Edaphological approach).

Pedological Approach: In pedology, the origin of the soil, its classification and its description are examined (from Greek word "pedon", means "soil" or "earth"). V. V. Dokuchaev defined pedology as "the science dealing with the laws of origin, formation, and geographic distribution of the soil as a body in nature". The study of soil as a natural body is Pedology and it does not focus on the soil's immediate practical use. Pedologists study, examine and classify soils as they occur in their natural environment. Their findings may be as useful to construction and highway engineers as to the farmers.

Edaphological approach: The study of soil from the standpoint of higher plants is Edaphology (from Greek word "edaphos", also means "soil" or "ground"). Edaphologists examine the various properties of soil in relation to crop production. Their approach has practical utility as they have the production of food and fibre as their ultimate goal. They determine the reasons for variations in the productivity of soils and find means for improvement.

1.3. Volume Composition of Soil

Soil is generally referred to as a *three-phase system* because it consists of solid (mineral and organic matter), liquid (water), and gas (air). The space occupied by water and air is called *void*. A saturated soil may be assumed to have only two of these phases, as voids are completely saturated with water.

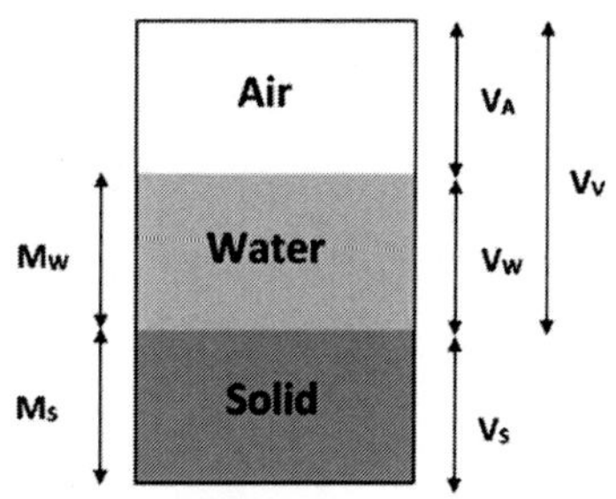

Fig. 1.1: The three-phase system of soil

The volume composition of a typical mineral soil is as follows:

1. Solid matrix:

   Mineral matter – 45%

   Organic matter – 5%

2. Pore space

   Soil water – 25%

   Soil air – 25%

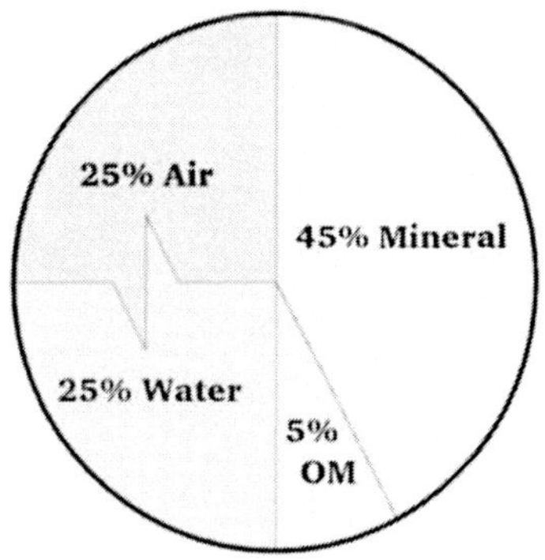

The space occupied by soil water and soil air are mutually exclusive.

2

Origin of Earth

During the 19th century various hypotheses were advanced about the origin of earth. These hypotheses can be grouped into two general classes:

1. *Monistic hypotheses*: The pioneers of this hypotheses believed that only one mass or body was responsible for the creation of the solar system. The representative of this theory is *Nebular hypothesis*. The advocates of this hypothesis were Kant and Laplace.
2. *Dualistic hypotheses*: The advocates of this hypothesis believed that two separate masses or bodies were responsible for the creation of the solar system. The representatives of this hypothesis are *Planetesimal hypothesis* and *Tidal hypothesis*. The pioneer of Planetesimal hypothesis was the great English scientist T. C. Chamberlain. The Tidal hypothesis was advocated by another English geographer named James Jeans.

2.1. The Nebular Hypothesis of Laplace

In the 18th century, the nebular hypothesis was first developed by Emanuel Swedenborg, Immanuel Kant, and Pierre-Simon Laplace. Kant argued that gaseous clouds - nebulae, which slowly rotated, gradually collapsed and flattened due to gravity and eventually formed stars and planets. Pierre Simon Marquis de Laplace (1749-1827), a French astronomer and mathematician extensively explained and modified Kant's hypothesis in 1785.

According to this hypothesis, the whole solar system is supposed to have existed at one time as a nebula or a cloudy mass of matter in the form of vapour or gases. As such it was intensely hot and incandescent. It was also supposed to be perpetually revolving. With the lapse of time, the nebula began to cool. On cooling, the outer layers of the vapours condensed and were thrown out in the form of rings. In this way, different rings were thrown out at different times. These rings, on further condensation and disruption, became what we know as the planets. The planets themselves, on cooling, threw off their own rings as is seen at present in the case of Saturn. In some cases, the planet rings broke up in their turn and formed satellites like moon. The central mass which is still incandescent and not yet condensed is the Sun.

The Nebular theory which was widely accepted for hundreds of years was rejected in the 20th century and planetesimal theory became popular. The major objection to the nebular theory was that rings cannot condense into planets. The most serious concerns the speed of rotation of sun. When this theory was worked out mathematically on the basis of known orbital momentum of planets, it predicted that Sun must rotate about 50 times more rapidly than it actually does.

2.2. Planetesimal Hypothesis of Chamberlain and Moulton

Planetesimal, a theory of origin of the solar system was proposed in 1905 by Forest Ray Moulton and Thomas C. Chamberlain, both of University of Chicago.

According to Chamberlain, initially there were two heavenly bodies (stars) in the Universe, *viz.*, Proto-sun and its companion star. The Proto-sun was formed of very small particles which were cold and solid. The introducing star came very close to the Proto-sun and infinite number of small particles was detached from outer surface of the Proto-sun due to the massive gravitational pull of this giant star. Chamberlain termed these particles as "planetesimals" [Fig. 2.1].

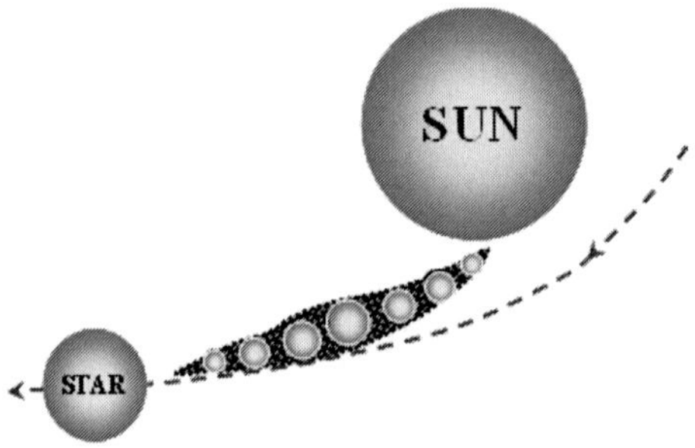

Fig. 2.1: Formation of planetesimals

Gradually, large planetesimals started attracting small ones and ultimately, they grew in the form of planets due to continuous accretion of infinite number of planetesimals. The remaining Proto-sun changed into the present-day Sun. The satellites of the planets were created due to the repetition of the same processes and mechanisms. According to this theory, the main force for ejection of small planetesimals was the tidal force exerted by the intruding (approaching) star on the outer surface of the Proto-sun.

The planetesimal hypothesis which was widely accepted for about 35 years later drew criticism for its drawbacks. The flaw in this theory was the assumption that the material drawn out of the stars would condense. The extremely hot gases that make up a star are held together by the gravitational forces within the star. The material would expand because of its heat, once it was pulled away to where the gravitational forces were weaker. Before condensation could take place, the gases would have almost entirely dissipated.

According to the planetesimal hypothesis, the nucleus of any orbit would have acquired and accreted any number of planetesimals and would have gained any size. Thus, the planets should not have been arranged in any order such as the present cigar-shaped solar system. The increase in size of the planets away from the Sun continues up to Jupiter and then the size goes on decreasing up to the last planet Pluto, the smallest and the farthest. Chamberlain does not offer any explanation to this type of arrangement. Another objection is that such a close encounter between the stars is not possible given the infinite space of the Universe.

During the 20th century, this model was challenged by a number of theorists who attempted to replace it with numerous models. However, none of these attempts were successful until the 1970's with Soviet astronomer Victor Safronov that the modern (widely accepted) Solar Nebular Disk (SND) Model came into being.

According to the SND model, our star system was formed 4.568 billion years ago when a small part of a giant molecular cloud experienced a gravitational collapse. While most of the collapsing mass collected in the centre formed the Sun, the rest flattened into a protoplanetary disk out of which the planets, moon, asteroids and other small solar system bodies formed. Since then, our system has evolved considerably due to collisions between objects, planetary migration and the capturing of extra-solar objects by our own system. Originally applied only to our own solar system, the SNDM has since been used by theorists to explain star formation throughout the known Universe.

Though this is the most widely accepted theory, alternative models do exist. Since the dawn of the space age in the 1950s and the discovery of extrasolar planets in the 1990s, all of these models have been both challenged and refined to account for new observations.

2.3. The Earth's Spheres

There are three spheres corresponding to three states of matter which constitute the earth. The solid zone is lithosphere; the covering of water, forming seas and oceans, is the hydrosphere and the gaseous envelop over the earth's surface is the atmosphere.

The atmosphere is composed of 78.08% nitrogen, 20.95% of oxygen, 0.93% of argon and 0.03% of carbon-di-oxide. In addition, inert gases such as helium, neon, krypton, xenon and radon are present. The water vapours present in the air vary in amounts at different places and times.

The hydrosphere covers almost three-fourths of the earth's surface. Most of it lies within the ocean basins; it also appears on the surface of land in the form of rivers, ponds, lakes and groundwater.

Lithosphere is the solid surface of continents, plains, plateaus and mountains, valleys, sand-dunes and lava flows and the interior of the earth that is composed of rocks and minerals. Lithosphere constitutes 93% of the earth's mass. It is covered by the gaseous envelop called the atmosphere.

2.4. The Interior of the Earth

The interior of the earth is divided into three concentric rings, *viz.*, *crust*, *mantle* and *core* (inner and outer). The *crust* varies in thickness from 5 to 11 km in the oceans and 25 to 40 km in the continents. It consists of rocks with density of 2.6 to 3.0 g cm^{-3}.

The *mantle* is the thick layer of solid rock below the crust. It is 2,900 km thick. The rock is composed of silicon, oxygen, aluminium, iron and magnesium with a density of 2.6 to 3.0 g cm^{-3}. The upper part of the mantle has a temperature of about 870 °C. This temperature gradually increases with depth through the mantle to about 4400 °C where the mantle meets the outer core.

The core is the inner most portion of earth. It is 3,500 km thick and is composed of metals such as iron and nickel, having an average density of 9.0 to 12.0 g cm^{-3}. The outer core is about 2,250 km thick, with temperature varying from 2,200 °C to 5,000 °C. The ball-shaped inner core is about 1,250 km thick, with a temperature of about 5,000 °C. While the density of the earth as a whole is 5.5 g cm^{-3}, that of its core is about 9.0 g cm^{-3}; the density increases from the crust to the core.

2.5. Composition of Earth's Crust

The earth's crust is mainly composed of mineral matter. The mineral matter is made up of various elements combined together to form compounds. Some elements exist in elemental form without forming compounds. Almost all the elements known to us are present in the earth's crust. Only a few, however, are present in significantly sufficient quantities [Table 2.1].

Table 2.1: Average composition of the earth's crust (% by weight)

Element	% by weight
Oxygen	46.60
Silicon	27.72
Aluminium	8.13
Iron	5.00
Calcium	3.63
Sodium	2.83
Potassium	2.59
Magnesium	2.09
Others	1.41

These elements do not exist in the earth's crust as such. Each element is in combination with one or more of other elements to form definite chemical compounds known as *minerals*. Many of these minerals in turn combine together to form aggregates called as *rocks*. Almost all the mineral matter is present in the form of rocks in earth's crust.

3

Soil Forming Rocks

Rock

A rock may be defined as "*an aggregate of one or more minerals*". An aggregate of single mineral calcite gives rise to crystalline limestone. Granite is a combination of several minerals. Most rocks are hard and compact as granite and basalt, but some are loose and feebly aggregated as sandstone. Rocks possess certain characteristics such as structure, colour, specific gravity, cleavage and mineralogical make-up by which we can identify the rocks and also distinguish one from the other.

3.1. Classification of Rocks

By origin or mode of formation, 3 great divisions of rocks are recognized:

- Igneous rocks
- Sedimentary rocks
- Metamorphic rocks

3.1.1. Igneous Rocks

Igneous (Latin) means "made from fire or heat". The earth's core contains liquid rock, called "*magma*". When volcanoes erupt, the magma from the Earth's core comes up to the earth's surface, gets cooled and new igneous rock is made. Igneous rocks form when molten magma cools and turns into solid rock. Thus, igneous rocks are the first formed rocks. These rocks are mostly crystalline and therefore are called '*crystalline rocks*'. They are also called '*massive rocks*' as they occur in masses. Igneous rocks consist mainly of primary minerals of which quartz, feldspars, amphiboles, pyroxenes and micas are most common.

Classification of Igneous Rocks

Based on the *mode of origin*, the igneous rocks are classified into two types:

1. Extrusive or Volcanic igneous rocks [Basalt, Dolerite, Dacite, Rhyolite, Pumice, Andesite]

2. Intrusive or Plutonic igneous rocks [Granite, Syenite, Diorite, Gabbro, Pegmatite, Peridotite]

Volcanic rocks are named after Vulcan, the Roman God of fire. These rocks are formed when the molten magma pours out on the surface of the earth where it gets consolidated on cooling. Because they are prone to cool quickly, they have a finer-grained matrix.

Intrusive rocks, named after Pluto, the Roman God of the underworld, are formed when the molten magma is cooled and consolidated beneath the earth's surface. Those intrusive rocks that are consolidated at a much greater depth than the others are called, '*plutonic rocks*'. Because the cooling process is slower, these rocks crystallize as coarse-grained. The intrusive rocks that get consolidated at relatively shallower depths are called, '*hypabyssal' or 'subvolcanic rocks'* and they are medium-grained.

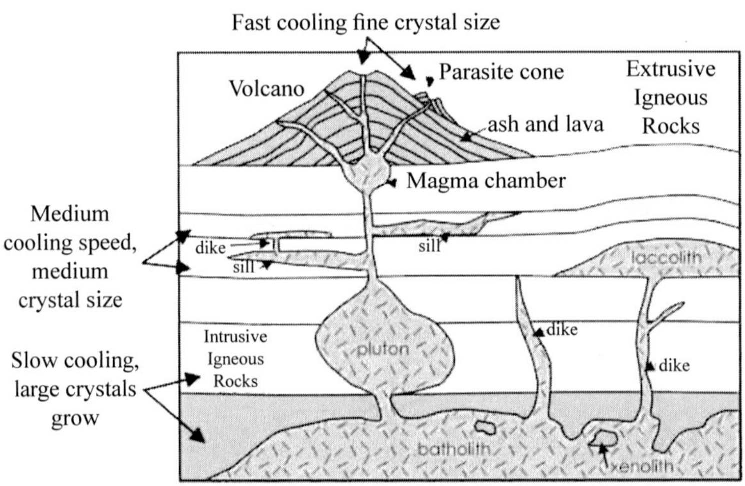

Fig. 3.1: Different types of igneous rocks

Other intrusive rocks were formed as a result of molten mass being forced among the pre-existing solid rocks through cracks and other planes of weakness. Some of these rocks consolidated in vertical cracks and formed wall-like masses are called, "*dykes*". Others that were consolidated into horizontal cracks or planes are known as, "*sills*" (Fig. 3.1).

Many igneous rocks of volcanic origin contain water under great pressure even when they are in molten condition. As the molten magma comes to the surface, pressure is released and water is at once turned into steam. As the steam cannot escape, it forms a globular cavity within the rock. The globular cavity is called "*vesicle*" and the rocks containing such vesicles are called, "*vesicular rocks*". The vesicles are very often filled with some mineral which has been introduced and deposited into the vesicle by infiltration of water-carrying those substances in solution. The rocks having these minerals embedded in its vesicle as though it is studded with almonds or amygdales. Hence such rocks are known as "*amygdaloidal rocks.*" The minerals so embedded are quartz, quartzite, zeolite, etc.

Based on *chemical composition* (more precisely, based on the SiO_2 content), the igneous rocks are classified into four types:

1. Acidic (65-75% SiO_2) [Granite, Rhyolite]
2. Neutral (55-65% SiO_2) [Andesite]
3. Basic (45-55% SiO_2) [Basalt, Diorite]
4. Ultra basic (<45% SiO_2) [Picrite, Gabbro, Limestone]

Igneous Rock	Acid	Intermediate	Basic
Volcanic	Rhyolite	Andesite	Basic
Plutonic	Granite	Syenite	Gabbro
Hypabyssal	Pegmatite	Microdiorite	Dolerite

3.1.2. Sedimentary Rocks

Sedimentary rocks cover 75% of the earth's surface. Most of the rocks found on the Earth's surface are sedimentary even though sedimentary rocks only make up less than 5% of all the rocks that make up Earth.

When rocks are exposed to such elements as air, rain, sun, freeze or thaw cycle, plants, erosion occurs. Tiny bits of rocks are worn away and get deposited as sediments. These sediments, with the passage of time, get buried by more and more sediments, become hardened and turn into sedimentary rocks. The sediments are often cemented together by substances like SiO_2, Fe_2O_3 or lime to give consolidated form.

Sedimentary rocks are usually formed in layers (strata) and hence they are also known as "*stratified rocks*". In some cases, these rocks are formed through chemical reactions, primarily as precipitates from aqueous solutions. Since these rocks are formed of substances which were at one time in solution, they are also called as "*aqueous rocks*".

In whatever way they are formed, sedimentary rocks are of secondary or derivative origin as they consist of materials that have previously existed on or beneath the surface of the earth and through transportation, accumulation and consolidation of these materials gave rise to new rocks. Therefore, sedimentary rocks consist mainly of secondary minerals together with some quantities of primary minerals.

Classification of Sedimentary Rocks

1. There are 6 main kinds of sedimentary rocks, *based on the appearance*:

Sedimentary Rock	Composition
Conglomerate	Pebbles, boulders cemented in a matrix
Sandstone	Sand grains cemented
Shale	Clay hardened into a rock
Limestone	Made of calcium carbonate
Gypsum, etc.	When seawater evaporates
Porphory	Jagged rock bits cemented

2. Sedimentary rocks are classified based on their *origin* as follows:

Type	Description	Example
Residual	Formed *in situ*	Laterite
	Deposited as solids in suspension	Sandstone, Shale
Transported	Deposited by chemical precipitation	Limestone, Iron stone
	Deposited through agency of organic matter	Peat, Phosphatic deposits

3. Based on the *grain size*, sedimentary rocks may be classified as follows:

   a. **Rudaceous**: Rocks consisting chiefly of boulders sand. [e.g. Conglomerate]

   b. **Arenaceous**: Rocks consisting chiefly of sand grade materials. [e.g. Sandstone]

   c. **Silt rocks**: Rocks consisting of silt-sized particles. [e.g. Silt stone]

   d. **Argillaceous**: Rocks consisting chiefly of small sized particles of clay. [e.g. Shale]

3.1.3. Metamorphic Rocks

Metamorphic rocks are the least common of the three kinds of rocks. Metamorphic rocks are rocks that have changed [Greek "meta" and "morph" "to change form"]. Originally, they were igneous or sedimentary, but have undergone metamorphism due to the influence of water, heat, pressure or chemically active liquids and gases. The changes so brought about are both physical and chemical in character, which alter the structural features in one case and the chemical composition in the other. The metamorphism so brought about by water is known as *hydro-metamorphism* and that by heat is *thermo-metamorphism*. The metamorphism brought about by pressure is called *dynamo-metamorphism*.

Sedimentary rock	Metamorphic form
Sandstone	Quartzite
Limestone	Marble
Shale	Slate / Mica schist
Clay	Schist
Granite	Granite gneiss

The inter-relationships between the three types of rocks are depicted in Fig. 3.2 below.

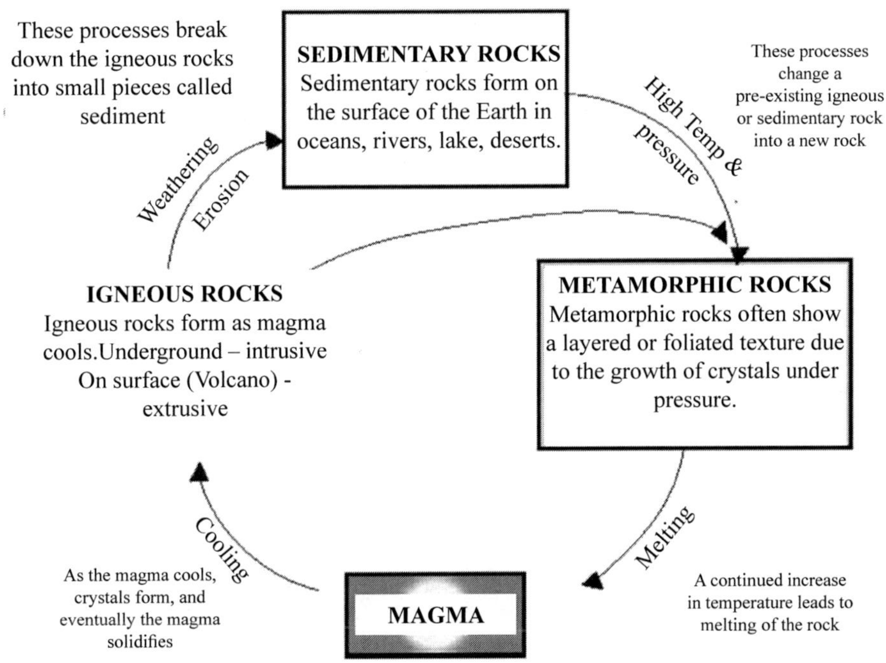

Fig. 3.2: Inter-relationships among the different types of rocks

3.2. Brief Description of Important Rocks

3.2.1. Igneous Rocks

Rock	Grain Size	Essential Minerals	Most common accessory mineral	Mean specific gravity	Remarks
Granite	Plutonic, crystalline	Predominantly quartz; 20-30% orthoclase	Hornblende Mica, Magnetite	2.64	Light coloured
Syenite	-do-	Predominantly quartz; 11% orthoclase	Hornblende, Biotite Magnetite	2.80	-do-
Granodiorite	-do-	Intermediate quartz; Plagioclase exceeds orthoclase	-do-	2.70	Medium coloured; reddish
Diorite	-do-	Intermediate quartz; Plagioclase absent	-do-	2.85	Darker
Gabbro	-do-	Augite + Olivine	Hornblende, Ilmenite	3.00	Blackish
Dolerite	Hypabyssal, ophitic texture	-do-	-do-	3.00	-do-
Basalt	Volcanic, Micro-crystalline with glassy mass	-do-	-do-	3.00	-do-

3.2.2. Sedimentary Rocks

Rock	Mineral composition	Colour & structure
Sandstone	Mainly quartz, with $CaCO_{3,}$ iron oxides and clays	Light to dark red; granular in structure
Shale	Clay minerals, quartz and some organic matter	Light to dark; thinly laminated structure
Limestone	Mainly calcite or calcite and dolomite with some iron oxides, clay, phosphate and organic matter	Usually light grey to yellow;

3.2.3. Metamorphic Rocks

Rock	Mineral composition	Colour & structure
Gneiss	Formed from granite; mineral composition as that of granite	Alternating light and dark colours; foliated structure
Schist	Formed from basalt or shale; mineral composition as that of original rock	Much as original rock; foliated structure
Quartzite	Formed from sandstone; of same composition	Light to brown; compact; non-foliated structure
Slate	Formed from shale; of same composition	Grey to black; compact; non-foliated structure
Marble	Formed from limestone; contains mainly calcite and dolomite with minor pigments as iron oxide	Light red, green, black; compact; fine to coarse texture; non-foliated structure

Granite (Coarse-grained)

Siltstone

Gneiss

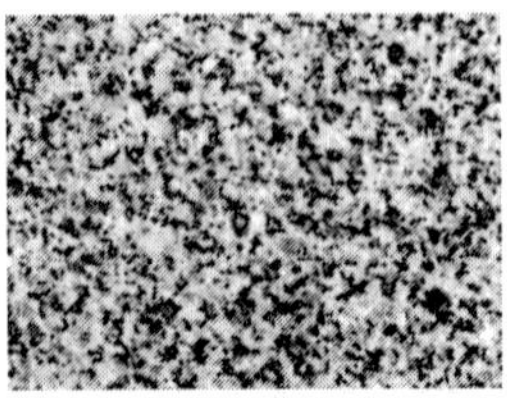
Diorite (Polished)

Limestone (Chalk)

Marble

Gabbro

Sandstone

Quartzite

Basalt

Shale

Slate

Plate 1: Some important rocks

4

Minerals

Mineral

A mineral is defined as *"a natural substance, usually inorganic, having a definite chemical composition and commonly a definite molecular arrangement which is usually expressed in geometric form"*.

4.1. Classification of Minerals

The soil forming minerals are classified into *primary minerals* and *secondary minerals*. The original mineral component of a rock is known as a primary mineral. The minerals formed due to the weathering of pre-existing primary minerals are called secondary minerals.

The primary minerals are further classified into *essential minerals* and *accessory minerals*. Those primary minerals which form the chief constituents of a rock and which are recognized as the characteristic components of that rock are called *essential minerals*. Those minerals which occur only in small quantities and whose presence is of no consequence as far as the characteristic of a rock is concerned are called *accessory minerals*. In other words, the essential minerals comprise the bulk of the rocks and the accessory minerals make up only a minor fraction of rocks.

4.1.1. Essential Minerals

The essential minerals are classified into:

1. Feldspars
   - 1.1. Potash feldspar [Orthoclase $KAlSi_3O_8$]
   - 1.2. Sodium feldspar [Albite $Na_2O.Al_2O_3.6SiO_2$]
   - 1.3. Calcium feldspar [Anorthite $CaO.Al_2O_3.2SiO_2$]
   - 1.4. Soda lime feldspar (Plagioclase) [Oligoclase]
2. Pyroxenes and amphiboles
3. Olivines (Glassy-appearing ferro magnesium silicate minerals)

Plate 2: Some important soil forming minerals

4. Mica

   4.1. Muscovite (White mica) $K_2O.3Al_2O_3.2H_2O$

   4.2. Biotite (Black mica) $H_2K(Mg.Fe)_3.Al(SiO_4)_3$

5. Zeolites (Hydrous Na, Ca, Al-silicates)

4.1.2. Accessory Minerals

The typical examples for accessory minerals are: apatite, pyrite (FeS_2), magnetite (Fe_3O_4), ilmenite ($FeO.TiO_2$) and rutile (TiO_2).

4.2. Classification of Minerals (By Chemical Composition)

The minerals can be classified into eight groups based on their chemical composition:

1. Native elements [Copper, silver, gold, nickel, iron, graphite, diamond]
2. Sulfides [Sphalerite, chalcopyrite, galena, pyrite]
3. Halides [Halite, fluorite]
4. Oxides and Hydroxides [Corundum, hematite]
5. Nitrates, Carbonates, Borates [Calcite, dolomite, malachite]
6. Sulfates, Chromates, Molybdates, Tungstates [Celestite, barite, gypsum]
7. Phosphates, Arsenates, Vanadates [Apatite, turquoise]
8. Silicates [Quartz, almandine garnet, topaz, jadeite, talc, biotite mica]

5

Weathering of Rocks and Minerals

Rocks and minerals formed under a very high temperature and pressure, when exposed to atmospheric conditions of low pressure and low temperature, they become unstable and weather. Soils are formed from rocks through the intermediate stage of formation of regolith (parent material) which is the resultant of weathering.

$$\text{Rocks} \xrightarrow{\text{Weathering}} \text{Parent Material} \xrightarrow{\text{Pedogenesis}} \text{Soil}$$

Weathering is a process of disintegration and decomposition of rocks and minerals which are brought about by physical agents and chemical processes, leading to the formation of regolith (unconsolidated residues of the weathering rock on the earth's surface or above the solid rocks). Weathering takes place by two basic processes - physical and chemical.

5.1. Physical Weathering

- Instrumental in cracking, crumbling, crushing and mellowing rocks and minerals into a mass of material varying in size from boulders to very fine particles
- Kind of mechanical disruption and is termed "disintegration"
- Change in shape, form, and size of rocks and minerals
- No change in chemical composition

The rocks are disintegrated and broken down to comparatively smaller pieces, without producing any new substances. The permeability of rocks is an important single factor. Coarse textured (porous) sandstone weathers more readily than a fine textured (almost solid) basalt. Unconsolidated volcanic ash weathers quicker than unconsolidated coarse deposits such as gravels. The agents of physical weathering are temperature, water, wind and glaciers.

5.1.1. Temperature

The variations in temperature exert great influence on the disintegration of rocks. During day time, the rocks get heated up by the sun and expand. At night, the temperature falls and the rocks get cooled and contract. This alternate expansion and contraction weaken the surface of the rocks and crumble it because the rocks do not conduct heat easily. Also, the minerals within the rock differ in their rate of expansion and contraction. For example, the expansion of quartz is twice as feldspar. As compared to light coloured rocks, dark coloured rocks are subjected to fast changes in temperature. Because the differential expansion of minerals in a rock surface generates stress between the heated surface and cooled unexpanded parts, it results in fragmentation of rocks. This causes the surface layer to peel off from the parent mass and the rock ultimately disintegrates. This process is called "*exfoliation*".

5.1.2. Water

Water acts as a disintegrating, transporting and depositing agent.

i) *Fragmentation and transport:* Rain water beats over the surface of the rock and starts flowing towards the ocean. Since moving water has the great cutting and carrying force, it forms gullies and ravines and carries with it suspended soil material of varying sizes. Transporting power of water varies as the sixth power of its velocity. Therefore, the greater the speed of water, greater is the transporting power and carrying capacity [Speed/sec carrying capacity: 15 cm for fine sand, 30 cm for gravel, 1.2 m for stones (1 kg) and 9.0 m for boulders of several tons]. The disintegration is greater near the source of river than near its mouth.

ii) *Action of freezing:* Frost has much more effect than heat in producing physical weathering. In cold regions, the water in the cracks and crevices freezes into ice and the volume increases to one tenth. Since freezing starts from the top there is no possibility of its upward expansion. Hence, the increase in volume creates enormous outward pressure which breaks apart the rocks.

iii) *Alternate wetting and drying:* Some natural substances increase considerably in volume on wetting and shrink on drying [Montmorillonite]. During dry weather, these clays shrink considerably forming deep cracks or wide cracks. On subsequent wetting, they swell. This alternate swelling and shrinking/ wetting or drying of clay enriched rocks make them loose and are eventually broken.

5.1.3. Wind

Wind has an erosive and transporting effect. When the wind is laden with fine materials such as fine sand, silt or clay particles, it has a serious abrasive effect. The sand laden winds itch the rocks and ultimately break down under force. The dust storm may transport tons of materials from one place to another. The shifting of soil causes serious wind erosion problem and may render cultivated land as degraded.

5.1.4. Glaciers

In cold regions, big glaciers start moving owing to the changes in temperature and/or gradient. While moving, they exert tremendous pressure over the rocks on which they pass and carry the loose materials. These materials get deposited on reaching the warmer regions, where its movement stops with the melting of ice.

5.2. Chemical Weathering

- Decomposes complex substances of rocks & minerals into simpler ones.
- Causes change in composition and is termed "decomposition".
- Some decomposition products go into solution, some split off as gases.
- Bulk remains in place.
- Parallel with decomposition, synthesis may take place.

Decomposition of rocks and minerals by various chemical processes is called *chemical weathering*. It is the most important process for soil formation. Chemical weathering takes place mainly at the surface of rocks and minerals with the disappearance of certain minerals and formation of secondary products (new materials).

Feldspar + water $\rightarrow$ clay mineral + soluble cations and anions

Chemical weathering becomes more effective with the increasing surface area of the rocks. The chemical reactions occur largely on the surface of the rocks; therefore, the smaller the fragments, the greater is the surface area per unit volume available for reaction. The effectiveness of chemical weathering is closely associated with the mineral composition of rocks. For example, quartz responds far slowly to the chemical attack than olivine or pyroxene. The following are the processes of chemical weathering.

5.2.1. Hydration

Hydration is chemical combination of water molecules with a particular substance or mineral leading to a change in structure. Soil forming minerals in rocks *per se* do not contain any water, but they undergo hydration when exposed to humid conditions. Upon hydration there is swelling and increase in volume of minerals. The minerals lose their lustre and become soft. It is a common process in nature and works with secondary minerals, such as aluminium oxide, iron oxide and gypsum.

$2Fe_2O_3 + 3\ H_2O \rightarrow 2Fe_2O_3.3H_2O$

(Haematite) (red) (Limonite) (yellow)

$Al_2O_3 + 3H_2O \rightarrow Al_2O_3.3H_2O$

(Bauxite) (Hydrated aluminium Oxide)

$CaSO_4 + 2H_2O \rightarrow CaSO_4\ .2H_2O$

(Anhydrite) (Gypsum)

$3(MgO.FeO.SiO_2) + 2H_2O \rightarrow 3MgO.2SiO_2.2H_2O + SiO_2 + 3H_2O$

(Olivine) (Serpentine)

5.2.2. Hydrolysis

Hydrolysis is the most important process in chemical weathering. Water dissociates into H^+ and OH^- ions which chemically combine with minerals and bring about changes such as exchange, decomposition of crystalline structure and formation of new compounds. Water acts as a weak acid on silicate minerals. Water containing CO_2 (absorbed from atmosphere), reacts with the minerals directly to produce insoluble clay minerals, positively charged metal ions (Ca^{++}, Mg^{++}, Na^+, K^+) and negatively charged ions (OH^-, HCO_3^-) and some soluble silica – all these ions are made available for plant growth.

$KAlSi_3O_8 + H_2O \rightarrow HAlSi_3O_8 + KOH$

(Orthoclase) (Silicic acid clay)

$HAlSi_3O_8 + 8\ H_2O \rightarrow Al_2O_3\ .3H_2O + 6\ H_2SiO_3$

(Recombination) (Hydrated aluminium oxide) (Silicic acid)

5.2.3. Solution

Some substances present in the rocks are directly soluble in water. Continuous action of water removes the soluble substances; the rock no longer remains solid and forms holes, rills or rough surface and ultimately falls into pieces or decomposes. The action is considerably increased when the water is acidified by the dissolution of organic and inorganic acids [Halites, NaCl].

$NaCl + H_2O \rightarrow Na^+, Cl^-, H_2O$ (dissolved ions with water)

5.2.4. Carbonation

Carbon-di-oxide when dissolved in water it forms carbonic acid.

$2H_2O + CO_2 \rightarrow H_2CO_3$

This carbonic acid attacks many rocks and minerals and brings them into solution. The carbonated water has an etching effect on some rocks, especially lime stone. The removal of the cementing materials that hold sand particles together leads to their disintegration.

$CaCO_3 + H_2CO_3 \rightarrow Ca(HCO_3)_2$

(Calcite slightly soluble) (Ca bicarbonate readily soluble)

5.2.5. Oxidation

Oxidation is the process of addition of oxygen to minerals. Usually, the absorption is from O_2 dissolved in soil water and that present in atmosphere. The oxidation is more active in the presence of moisture and results in hydrated oxides [Minerals containing Fe and Mg].

$4FeO$ (Ferrous oxide) + $O_2 \rightarrow 2Fe_2O_3$ (Ferric oxide)

$4Fe_3O_4$ (Magnetite) + $O_2 \rightarrow 6Fe_2O_3$ (Haematite)

$2Fe_2O_3$ (Haematite) + $3H_2O \rightarrow 2Fe_2O_3.3H_2O$ (Limonite)

5.2.6. Reduction

Reduction, the process of removal of oxygen, is equally important in changing soil colour to grey, blue or green as ferric iron is converted to ferrous iron compounds. Under the conditions of excess water (less or no oxygen), reduction takes place.

$2Fe_2O_3$ (Haematite) $\rightarrow 4FeO$ (Ferrous oxide) + $O_2\uparrow$

5.3. Biological Weathering

Unlike physical and chemical weathering, the biological agents are responsible for both decomposition and disintegration of rocks and minerals. The biological life is controlled largely by the prevailing environment.

5.3.1. Man and Animals

The action of man in disintegration of rocks is well known as he cuts rocks to build dams, roads and buildings. All these activities result in increasing the surface area of the rocks for attack of chemical agents and accelerate the process of decomposition. A large number of animals, birds, insects and worms, by their activities aid in weathering. In tropical and sub-tropical regions, ants and termites build galleries and passages ("*Crotovina*") and carry materials from lower to upper surface and excrete acids. The oxygen and water with many dissolved substances, reach every part of the rock through the cracks, holes and galleries, and bring about speedy disintegration. Rabbits, by burrowing into the ground, destroy soft rocks. Moles, ants and bodies of the dead animals, provide substances which react with minerals and aid in decaying process. The earthworms pass the soil through the alimentary canal and thus bring about physical and chemical changes in soil material.

5.3.2. Higher Plants and Roots

The roots of trees and plants penetrate into the joints and crevices of the rocks. As they grow, they exert a great disruptive force and the rocks may break apart. The grass roots form a sponge-like mass, prevent erosion and conserve moisture, thus allowing moisture and air to enter into the rock for further action. Some roots penetrate deep into the soil and may open some sort of drainage channel. The roots running in crevices in limestone and marble produce acids. These acids have a solvent action on carbonates. The dead roots and plant residues decompose and produce organic acids which are of great importance in weathering.

5.3.3. Micro-organisms

In early stages of mineral decomposition and soil formation, the lower forms of plants and animals like mosses, bacteria, fungi and actinomycetes play an important role. Extracting nutrients from the rock and N from air, they live with a small quantity of water. In the due course of time, soil develops under the cluster of these micro-organisms. These organisms are closely associated with the decay of plant and animal remains and thus liberate nutrients for the use of next generation plants and also produce CO_2 and organic compounds which aid in mineral decomposition.

5.4. Weathering of Minerals

There are many factors which influence the weathering of minerals: Climatic conditions, physical characteristics, chemical and structural characteristics.

5.4.1. Climatic Conditions

The climatic condition, more than any other factor, tends to control the kind and rate of weathering. Under conditions of low rainfall, there is a dominance of physical weathering which reduces the size and increases the surface area with little change in volume. The increase in moisture content encourages chemical changes and new minerals and soluble products are formed. The weathering rates are generally fastest in the humid tropics as there is sufficient moisture and warmth to encourage chemical decomposition. While the easily weatherable minerals disappear on account of intense chemical weathering, more resistant products (hydrous oxides of Fe and Al) accumulate. Climate controls the dominant type of vegetation which in turn controls the mineral weathering.

5.4.2. Chemical and Structural Characteristics

Chemical and crystalline characteristics determine the ease of decomposition of minerals of given particle size. Ferro magnesium minerals are more susceptible to chemical weathering than feldspar and quartz. Less tightly packed minerals like olivine and biotite are easily weathered as compared to tightly packed zircon and muscovite (resistant).

5.4.3. Weathering Sequence

The minerals that remain in the soil and regolith have led to the following generalized order of weathering:

Quartz (most resistant) > muscovite > K feldspar > Na and Ca feldspar > biotite, hornblende, augite > olivine > dolomite and calcite > gypsum (least resistant)

6

Soil Forming Factors

Soil formation takes place in two consecutive stages:

1. The weathering of rocks into regolith
2. The formation of true soil from regolith

The evolution of true soil from regolith takes place by thc combined action of soil forming factors and processes. The first step is accomplished by weathering (disintegration and decomposition) and the second is associated with the action of soil forming factors which are known as pedogenic factors.

6.1. Soil Forming Factors

Dokuchaev (1889) established that the soils develop as a result of the action of soil forming factors:

$$S = f(p, Cl, o)$$

Further, Jenny (1961) formulated the following equations:

$S = f$ (climate) o, r, p, t … Climofunction

$S = f$ (organisms) cl, r, p, t … Biofunction

$S = f$ (topography) cl, o, p, t … Topofunction

$S = f$ (parent material) cl, o, r, t … Lithofunction

$S = f$ (time) cl, o, r, p … Chronofunction

where S = any soil property as OM, clay content, etc.

f = a function of

Cl= climate

o = organism

r = relief

p = parent material

t = time

... = additional, unspecified factors as fires, etc.

6.1.1. Active Factors

The active factors of soil formation are climate and organisms.

Climate

The most significant factor controlling the type and rate of soil formation is climate. The dominant climates recognized are:

Arid climate: The precipitation here is far less than the water-need and, therefore, the soils remain dry for most of the time in a year.

Humid climate: The precipitation is much more than the water-need and the excess water results in leaching of salt and bases, followed by translocation of clay colloids.

Oceanic climate: Moderate seasonal variation of rainfall and temperature.

Mediterranean climate: Moderate precipitation. Winters and summers are dry and hot.

Continental climate: Warm summers and extremely cool or cold winters.

Temperate climate: Cold humid conditions with warm summers.

Tropical and subtropical climate: Warm to hot humid with isothermal conditions.

Climate affects soil formation directly and indirectly. Directly, climate affects soil formation by supplying water and heat to react with parent material. Indirectly, it determines the fauna and flora activities that furnish a source of energy in the form of organic matter. This energy acts on rocks and minerals in the form of acids and salts are released. The indirect effects of climate on soil formation are most clearly evident in the relationship of soils to vegetation. Precipitation and temperature are the two major climatic elements that contribute most to soil formation.

Precipitation

Precipitation is the most important among the climatic factors. As it percolates and moves from one part of parent material to another it carries substances in solution as well as in suspension. The substances so carried are redeposited in another part or completely removed from the material when the soil moisture at the surface evaporates causing an upward movement of water. The soluble substances move with it and are translocated to the upper layer. Thus, a redistribution of substances, both soluble as well as in suspension in soil body, is brought about by rainfall.

Temperature

High temperatures hinder leaching, cause an upward movement of soluble salts, favor rapid decomposition of organic matter and increase microbial activities in soil. On the other hand, low temperatures induce leaching by reducing evaporation and favour the accumulation of organic matter by slowing down the processes of decomposition. Thus, the rate of chemical and biological reactions taking place in the parent material is controlled by temperature. According to Jenny (1941), the rate of weathering in the tropical regions proceeds three times faster than in temperate regions and nine times faster than in the arctic.

Organisms and Vegetation

The active components of soil ecosystem are microorganisms, plants, animals and man. Microorganisms have a role in soil formation that is related to the humification and mineralization of vegetation.

Burrowing animals, dig and mix-up the soil mass and thus disturb the parent material. Through his manipulation of natural vegetation, agricultural practices, etc., man influences soil formation. Compaction by traffic of man and animals decreases the rate of water infiltration into the soil, thereby increasing the rate of runoff and erosion. Rodents carry materials into their nests (usually B horizon) from upper horizon and bring up 'B' material into 'A' horizon. Rodents (mice, moles, ground squirrels, crayfish, etc.) make deep burrows, with complicated tunnels and passage ways. Plant roots follow abandoned tunnel borings.

Earthworms penetrate soil depths from few inches to 25 feet. Generally, they are active in soils rich in organic matter. Over 150,000 earthworm burrows have been reported in one acre. Materials passing through bodies of earthworms undergo chemical and physical changes. Ants, termites carry material from lower to higher levels, build galleries, underground passage ways, form hills & mounds. "*Crotovina*" is a term (derived from Russian) used to denote passage ways and nests of rodents.

Plants roots penetrate into the parent material and act both mechanically and chemically. They facilitate percolation and drainage and bring about greater dissolution of minerals through the action of CO_2 and acidic substances secreted by them. The decomposition and humification of the materials further add to the solubilization of minerals. Forests reduce temperature and evaporation and increase humidity and precipitation. Grasses reduce runoff and result in greater penetration of water into the parent material.

In summary, soil development proceeds at a rate that is a combination of the effects of time plus the intensities of climate and biota (organisms), further modified by the effect of land relief (topography) on which the soil is situated and the kind of parent material from which it is developing.

6.1.2. Passive Factors

The passive soil forming factors are parent material, relief and time. These provide a base on which the active soil forming factors work for the development of soil.

Parent Material

It is that mass (consolidated material) from which the soil has formed. A soil "inherits" its traits from the parent material from which it formed. Thus, soils that form from limestone are rich in calcium and soils that form from materials at the bottom of lakes are high in clay. Every soil is formed from parent material deposited at the Earth's surface. The material could be the bedrock that weathered *in situ* or smaller materials carried by flooding rivers, moving glaciers or blowing winds. Parent material is changed through biological, chemical and environmental processes, such as weathering and erosion. Different parent materials give rise to same kind of soil, if principal factors (climate and organisms) are similar. Similar parent materials give rise to different kinds of soil, if principal factors are dissimilar. Parent materials influence soil texture, soil chemistry, and nutrient cycling. Soils brought about by means of properties of parent materials rather than of external conditions are called "*endodynamomorphic soils*".

Origin of PM	Agent	Name
Sedentary	Still at original site	Residual
Transported	Gravity	Colluvial
	Water	Alluvial deposit Marine deposit Lacustrine
	Ice	Glacial
	Wind	Aeolian (sand) Loess (silt)

Ectodynamomorphic soils: Development of normal profile under the influence of climate and vegetation. The soils developed on transported parent materials bear the name of the parent material; *viz*. alluvial soils from alluvium, colluvial soils from colluviums, etc.

Different parent materials influence profile development and produce different soils, especially in the initial stages. Acid igneous rocks, like granite and rhyolite, produce light-textured soils (Alfisols). Basic igneous rocks (such as basalt), alluvium or colluvium derived from limestone or basalt, produce fine-textured cracking-clay soils (Vertisols). Basic alluvium or aeolian materials produce fine to coarse-textured soils (Entisols or Inceptisols). The kind of the elements released during the decaying of rocks has a specific role in soil formation. Silicon and aluminium form the skeleton for the production of secondary clay minerals. Iron and manganese are important for imparting red colour to soils and for oxidation and reduction phenomena. Calcium and magnesium have a flocculating effect, resulting in favorable and stable soil structure for plant growth.

Relief

Relief and topography are used as synonyms. They denote the configuration of the land surface. The shape of the land and the direction it faces make a difference in how much sunlight the soil gets and how much water it keeps. Because gravity and water move soil particles down the slope, deeper soils form at the bottom of a hill. The prominent types of topography designations, as per FAO Guidelines (1990) are:

- Flat to almost flat: 0 – 2 %
- Gently undulating: 2 - 5 % slope
- Undulating: 5 – 10 %
- Rolling: 10 – 15 %
- Hilly: 15 –3 0 %
- Steeply dissect: > 30 % with moderate range of elevation (<300 m)
- Mountainous: > 30% with great range of elevation (>300 m)

On a level topography, almost the entire water received through rain percolates through the soil. In such conditions, the soils formed may be considered to represent the regional climate. They have normal solum with distinct horizons. But vast and monotonous level land with little gradient often impairs drainage conditions.

Due to accelerated erosion, which removes surface material before it has the time to develop, the soils on steep slopes are generally shallow, stony and have weakly-developed profiles with less distinct horizonation.

In semi-arid and sub-humid regions, the depression areas reflect more moist conditions than actually observed on level topographic positions due to the additional water received as runoff. Such conditions (as in the Tarai region of the Uttar Pradesh) favour more vegetative growth and slower rate of decay of organic remains. This results in the formation of comparatively dark-coloured soils rich in organic matter (Mollisols).

Topography affects soil formation by influencing temperature and vegetative growth through slope exposures (aspect). The southern exposures (facing the sun) are warmer and are subject to marked fluctuations in temperature and moisture. The northern exposures, on the other hand, are much cooler and humid. The eastern and western exposures occupy intermediate position in this respect.

Time

Soil formation, being a very slow process, requires thousands of years to develop a mature pedon. The duration taken by a given soil to advance from the stage of weathered rock (i.e. regolith) to the stage of maturity is considered as time of soil formation. All of factors of soil formation work together over time. Older soils differ from younger ones because they have had longer time to develop. As soil ages, it starts to look different from its parent material and this is because soil is dynamic. All its components - minerals, water, air, organic matter, and organisms - constantly change. Components are added and lost. Some components move from place to place within the soil and some are totally changed, or transformed. It takes hundreds of years to develop an inch of soil. The time that nature devotes to the formation of soils is termed as 'pedologic time'. A soil with fully developed horizons (A, B, C) denotes it is mature.

It has been observed that rocks and minerals disintegrate and/or decompose at different rates; the coarse particles of limestone are more resistant to disintegration than those of sandstone. However, in general, limestone decomposes more readily than sandstone (by chemical weathering).

The following are the weathering stages in soil formation:

1. Initial [Unweathered parent material]
2. Juvenile [Weathering started but much of the original material still unweathered]
3. Virile [Easily weatherable minerals fairly decomposed; clay content increased, slowly weatherable minerals still appreciable]
4. Senile [Decomposition reaches final stage; only most resistant minerals survive]
5. Final [Soil development completed under prevailing environments]

The soil properties also change with time; for example, nitrogen and organic matter contents increase with time if the soil temperature is not high. The $CaCO_3$ content may decrease or even be lost with time, provided the climatic conditions are not arid. In humid regions, the H^+ concentration increases with time because of chemical weathering.

7

Soil Forming Processes

The pedogenic processes are dynamic and extremely complex involving several chemical and biological reactions, usually operating simultaneously in a given area. One process may counteract another, or two different processes may work in tandem to achieve the same result. Different processes or combination of processes operate under varying natural environments. The combined interaction of various soil forming factors under different environmental conditions set a furtherance to certain recognized soil forming processes. Pedogenic processes include:

1. Additions of organic and mineral materials to a soil body (solids, liquids, gases)
2. Losses of these from a soil body
3. Translocation of materials from one point to another within a soil body
4. Transformation of mineral and organic substances within a soil body

Pedogenic Process	Type	Description
Eluviation	3	Movement of material out of a portion of a soil profile
Illuviation	3	Movement of material into a portion of soil profile
Leaching	2	Washing out of soluble materials from the solum
Enrichment	1	Addition of material to a soil body
Erosion	2	Removal of material from surface layer of soil
Cumulization	1	Aeolian & hydrologic additions of mineral particles to surface layer
Salinization	3	Accumulation of soluble salts as chlorides, sulphates, etc.
Desalinization	3	Removal of soluble salts
Solonization	3	Accumulation of Na^+ ions on exchange sites in a soil
Solodization	3	Leaching of Na^+ ions from soil horizon
Humification	4	Transformation of raw organic materials into humus
Mineralization	4	Release of minerals through decomposition of org. matter
Lessivage	3	Mechanical migration of clay and silt particles from A to B horizon
Pedoturbation	3	Biological, physical churning of soil materials, homogenizing solum in varying degrees

7.1. Fundamental Soil Forming Processes

7.1.1. Humification

Humification is the extremely complex process of transformation of raw organic matter into humus involving various organisms. First, simple compounds such as sugars and starches are attacked, followed by proteins and cellulose and finally very resistant compounds, such as tannins, are decomposed culminating in the formation of dark coloured humus.

7.1.2. Eluviation

It is the mobilization and translocation of constituents as Clay, Fe_2O_3, Al_2O_3, SiO_2, humus, $CaCO_3$, etc. from one point of soil body to another. Eluviation is the process of removal of constituents in suspension or solution by the percolating water from the upper to lower layers. Eluviation includes mobilization and translocation of mobile constituents leading to textural differences. The horizon formed by the process of eluviation is termed as eluvial horizon (A_2 or E horizon).

7.1.3. Illuviation

The process of deposition of soil materials removed from the eluvial horizon in the lower layer is termed as illuviation. The horizons formed by this process are termed as illuvial horizons (B-horizons, especially B_t).

7.1.4. Horizonization

Two overlapping trends in soil development are horizonization and haploidization. Horizonization denotes processes and conditions by which initial materials are differentiated into profiles with many horizons.

7.1.5. Haploidization

Haploidization includes processes and conditions by which horizonization is inhibited or by which horizons are mixed.

7.2. Specific Soil Forming Processes

The basic pedologic processes provide a framework for later operation of more specific processes.

7.2.1. Calcification

It is the process of precipitation and accumulation of calcium carbonate in some part of the profile. The accumulation of $CaCO_3$ may result in the development of a calcic horizon. Calcium is readily soluble in acidic soil water or when CO_2 concentration is high in root zone as:

$CO_2 + H_2O \rightarrow H_2CO_3$

$4H_2CO_3 + 2Ca \rightarrow 2\ Ca(HCO_3)_2$ (Soluble) + 2H

$Ca(HCO_3)_2 \rightarrow CaCO_3 \downarrow + H_2O + CO_2 \uparrow$

The process of precipitation after mobilization under these conditions is called calcification and the resulting illuviated horizon of carbonates is designated as B_k horizon (B_{Ca}).

7.2.2. Decalcification

It is the reverse of calcification that is the process of removal of $CaCO_3$ or calcium ions from the soil by leaching.

$CaCO_3 + CO_2 + H_2O \rightarrow Ca(HCO_3)_2$

(Insoluble) (Soluble)

7.2.3. Podzolization

It is a process of soil formation leading to the formation of podzols and podzolic soils. In many aspects, podzolization is the negative of calcification and is essentially one of acid-leaching. The calcification process concentrates calcium in the lower part of the B horizon, while podzolization leaches calcium carbonates from the entire solum. Apart from calcium, the other bases are also removed and because of this the whole soil becomes distinctly acidic. In fact, the process operates under favorable combination of the following environments:

i) *Climate*: A cold and humid climate is most favorable.

ii) *Parent material*: Siliceous (sandy) material, having poor reserves of weatherable minerals, favour the operation of podzolization as it helps in easy percolation of water.

iii) *Vegetation*: Acid producing vegetation such as coniferous pines are essential.

iv) *Leaching and translocation of sesquioxide*: In the process of decomposition of organic matter various organic acids are produced. The organic acids thus formed act with sesquioxide and the remaining clay minerals, forming organic-sesquioxide and organic clay complexes, which are soluble and move with the percolating water to the lower horizons (B_h, B_s). Aluminium ions in a water solution hydrolyze and make the soil solution very acidic.

$2Al + 6H_2O \rightarrow 2\ Al(OH)_3 + 6H^+$

As iron and aluminium move out, the A horizon gets a bleached grey or ashy appearance. The Russians used the term 'podzols' ('pod' means under, the 'zola' means ash-like) for such soils due to the ash-like horizon appearing beneath the surface horizon. Thus, podzolization is a soil forming process which prevails in a cold and humid climate where coniferous and acid forming vegetations dominate. The humus and sesquioxide are leached out from the upper horizon and deposited in the lower horizons.

7.2.4. Laterization

The term 'laterite', derived from the word 'later' meaning 'brick' or 'tile', originally denoted a group of Indian soils in Malabar hills of Kerala, Tamil Nadu, Karnataka and Maharashtra that had a unique cemented horizon which when dried, becomes very hard, like a brick. Such soils (in tropics) when massively impregnated with sesquioxides (iron and aluminium oxides) to the extent of 70 to 80 per cent of the total mass are called "*laterites*" or "*latosols*" (Oxisols). The soil forming process is called "*laterization*".

Laterization is a process that removes silica, instead of sesquioxides, from the upper layers, thereby leaving sesquioxides to concentrate in the solum. The process operates under the following conditions.

i) *Climate*: The process of laterization operates most favorable in warm and humid (tropical) climate with 2000 to 2500 mm rainfall and continuous high temperature (25 °C) throughout the year.

ii) *Natural vegetation*: The rain forests of tropical areas are favorable for the process.

iii) *Parent material*: Basic parent materials with sufficient iron-bearing ferromagnesium minerals (pyroxene, amphiboles, biotites and chlorites), which upon weathering release iron, are congenial for the development of laterites.

7.2.5. Gleization

The term 'glei' of Russian origin means blue, grey or green clay. It is a process of soil formation resulting in the development of a glei (gley horizon) in the lower part of the soil profile above the parent material due to poor drainage conditions (lack of oxygen). Such soils are "hydromorphic soils". The process is not particularly dependent on climate (high rainfall as in humid regions) but often on drainage conditions. Poor drainage conditions result from lower topographic position, such as depression land, where water stands continuously at or close to the surface, impervious soil parent material and lack of aeration. Under such conditions, iron compounds are reduced to soluble ferrous forms. The solubility of Ca, Mg, Fe, and Mn is increased and most of the iron exists as Fe^{++} in solution or as mixed precipitate of ferric and ferrous hydroxides. This is responsible for the production of typical bluish to grayish horizon with mottling of yellow and or reddish brown colors.

7.2.6. Salinization

Salinization is the process of accumulation of salts in soils, whereby salts such as sulphates and chlorides of calcium, magnesium, sodium and potassium get accumulated in the form of a salty (salic) horizon. It is quite common in arid and semi-arid regions. It may also take place through capillary rise of saline ground water and by inundation of seawater in marine and coastal soils.

7.2.7. Desalinization

It is the removal by leaching of excess soluble salts from horizons or soil profile (that contained enough soluble salts to impair the plant growth) by ponding water and improving the drainage conditions by installing artificial drainage network.

7.2.8. Solonization or Alkalization

The process involves the accumulation of sodium ions on the exchange complex of the clay, resulting in the formation of sodic soils (solonetz). All cations in solution are engaged in a reversible reaction with the exchange sites on the clay and organic matter particles.

7.2.9. Solodization or Dealkalization

The process refers to the removal of Na^+ from the exchange sites. This process involves dispersion of clay. Dispersion occurs when Na^+ ions become hydrated. Much of the dispersion can be eliminated if Ca^{++} and or Mg^{++} ions are concentrated in the water. Ca^{++} and Mg^{++} ions can replace the Na^+ on exchange complex, and the salts of sodium are leached out as:

$$2NaX + CaSO_4 \rightarrow Na_2SO_4 + CaX$$

7.2.10. Pedoturbation

Pedoturbation is the process of mixing of the soil. Mixing to a certain extent takes place in all soils. The most common types of pedoturbation are:

Faunal pedoturbation: It is the mixing of soil by animals such as ants, earthworms, moles, rodents, and man himself.

Floral pedoturbation: It is the mixing of soil by plants as in tree tipping that forms pits and mounds.

Argillic pedoturbation: It is the mixing of materials in the solum by the churning process caused by swell shrink clays as observed in deep black cotton soils.

8

Soil Profile

There are over 100,000 different types of soil in the world, each with its own set of characteristics. If any soil is dug deep, one can see that it is made of layers, or horizons. The horizons together form a soil profile. Each profile tells a story about the life of a soil.

The development of soil profile is a constructive process where in disintegrated materials resulting from weathering of rocks and minerals get converted into a soil body. *Soil Individual or Polypedon* is a natural unit of soil that differs from its adjoining unit on the landscape in one or more properties. In modern soil taxonomy, polypedon is the unit of soil mapping. A *pedon* is the smallest volume recognizable in nature that can be called "*a soil*".

8.1. Soil Profile Defined

The exposed vertical section of the soil showing the various layers from the surface to the unaffected parent material is known as a soil profile.

8.2. Horizons

A soil profile exhibits three main horizons, named as horizon A, horizon B and horizon C. The surface soil or that layer of soil at the top which is prone to leaching and from which some soil constituents have been removed is known as horizon A or the horizon of eluviation. The intermediate layer in which the materials leached from horizon A have been re-deposited is known as horizon B or the horizon of illuviation. The parent material from which the soil is formed is horizon C [Fig. 8.1].

A study of soil profile is important as it is a historic record of all the soil forming processes that has taken place over years and it forms the basis for pedological investigations. Soil profile is the key to soil classification and forms the basis for the practical utility of soils.

8.3. Master Horizons and Sub Horizons

O horizon - It is called as organic horizon and is formed in the upper part of the mineral soil, dominated by fresh or partly decomposed organic materials. This horizon contains more than 30% organic matter if mineral fraction has more than 50 % clay (or) more than 20 % organic matter if mineral fraction has less clay. The organic horizons are commonly seen in forest areas and generally absent in grassland and cultivated soils.

O_1 Organic horizon in which the original forms of the plant and animal residues can be recognized through naked eye.

O_2 Organic horizon in which the original plant or animal matter cannot be recognized through naked eye.

A horizon - Horizon of organic matter accumulation adjacent to surface and that has lost clay, iron and aluminium.

A_1 Top most mineral horizon formed adjacent to the surface. There will be accumulation of humified organic matter associated with mineral fraction and darker in colour than that of lower horizons due to organic matter.

A_2 **(E horizon)** - Horizon of maximum eluviation of clay, Fe and Al oxides and organic matter. Loss of these constituents generally results in accumulation of quartz and other sand and silt size resistant minerals. Generally lighter in colour than horizons above and below.

A_3 **(AB or EB)** - A transitional layer between A and B horizons with more dominated properties of A_1 or A_2 above than the underlying B horizon. This horizon is sometimes absent.

B horizon - Horizon in which the dominant features are accumulation of clay, iron, aluminium or humus alone or in combination. Coating of sesquioxides will impart darker, stronger of red colour than overlying or underlying horizons.

B_1 **(BA or BE)** - A transitional layer between A and B. More like A than B.

B_2 **(B or B_w)** - Zone of maximum accumulation of clay, Fe and Al oxide that may have moved down from upper horizons or may have formed in situ. The organic matter content is generally higher and colour darker than that of A_2 horizon above.

B_3 **(BC or CB)** - Transitional horizon between B and C, with properties more similar to that of overlying B_2 than the underlying C.

C horizon - It is the horizon below the solum (A + B), relatively less affected by soil forming processes. It is outside the zone of major biological activity. It may contain accumulation of carbonates or sulphates, calcium and magnesium

R Underlying consolidated bedrock and it may or may not be like the parent rock from which the solum is formed.

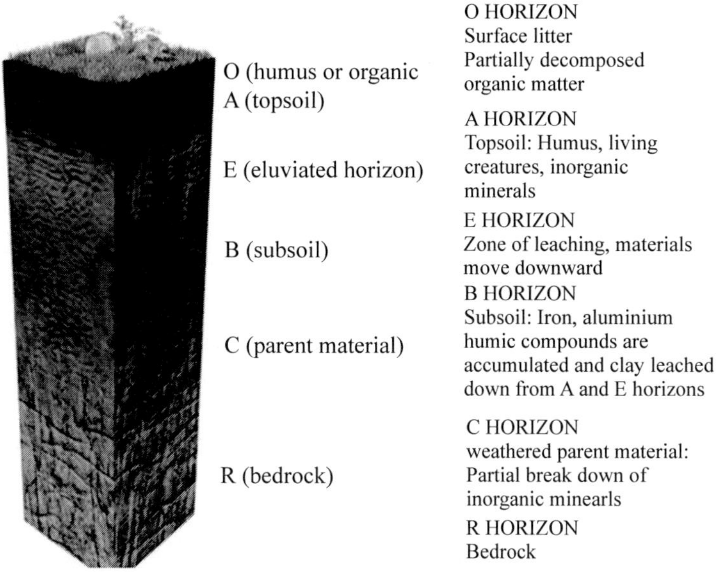

Fig. 8.1: A typical soil profile

A Mollisol profile showing thick dark A horizon with high organic matter content

An Oxisol profile (latosol) in Kerala, with no distinctive horizons and rich in iron oxides

A Vertisol profile with high content of montmorillonite clay

Plate 3: Some important soil profiles

Besides, lower case letters are used to indicate the special features of master horizons [Ap - ploughed layer; B_{2t} - illuvial clay]. When two or more genetically unrelated (contrasting) materials are present in a profile then the phenomenon is known as "*lithological discontinuity*". This is indicated by the use of Roman letters as prefixes to the master horizons [*e.g.*, A_p, B_2, II B_{22}, IIIC].

9

Soil Taxonomy

Soil classification, or soil taxonomy as it is called, is grouping of soils in an orderly and logical manner based on soil properties (differentiating characters) for the purpose of studying and identifying them. The following are the purpose of soil classification:

- To organize our knowledge of soils
- To bring out and understand the relationships among the soils
- To remember the properties of soils
- To learn new relationships and principles in different soil groups
- To establish groups or subdivisions or classes of soils
- To map the soils
- To predict the behavior of soils under defined use and management options

9.1. Brief History of Soil Classification

The first attempts of soil classification can perhaps be credited to the farmers who used straightforward criteria such as colour of the topsoil, consistency, permeability, risk for waterlogging, etc., to recognize soils, their potentials and constraints. Subsequently, soil scientists have done their very best through numerous individual nomenclatures. There are three linked development stages in the history of soil classification:

Stage	I	II	III
Emphasis	Environment →	Pedogenic Processes →	Soil characteristics
Criteria	Pedogenic factors such as climate, topography, etc.	Chemical and physical processes: Podzolization, salinization, leaching, etc.	Soil properties such as base saturation, clay mineralogy, etc.
Classification	Russian School (1890-1930)	W. European classifications (1940 - 1960)	Soil Taxonomy (1960 - 1990)

Early soil classification systems (Russian, mainly) focused on the environment and the soil forming factors to classify soils into zonal soils (fully developed profile, reflecting influence of climate and vegetation, as Boreal soil), azonal soils (poorly developed profile because of time as limiting factor, as alluvial soils and regosols) and intrazonal soils (soils occurring within the zone, but largely influenced by local conditions such as topography, parent material, as calcimorphic, hydromorphic soils). A later development focused on the processes occurring in the soil itself (such as salinization, leaching, accumulation, etc.). A good example of this approach is the French classification system (CPCS, 1967). The modern era of soil classification began with the publication of the 7th Approximation of the USDA Soil Taxonomy which used precisely defined and quantified soil properties to define "diagnostic soil horizons".

9.2. World Soil Classification Systems

There are several soil classification systems around the world and discussing all those are beyond the scope of this book. Therefore, only a few internationally accepted systems are narrated below.

9.2.1. FAO/UNESCO Legend of the Soil Map of the World

In order to prepare a soil map of the world as requested by the International Union of Soil Sciences (IUSS) in 1968, FAO developed a supra-national legend which has been used as an international soil classification system. The FAO/UNESCO Legend is a very simple, truly first international system and most soils could be accommodated on the basis of their field descriptions. The FAO Soil Map was designed for mapping soils at a continental scale. A major revision of the system was published in 1988. This system was finally replaced by the World Reference Base for Soil Resources (WRB) in 1998.

9.2.2. World Reference Base

The WRB is the international standard for soil classification system endorsed by the IUSS. Developed by an international collaboration synchronized by the IUSS Working Group, it replaced the FAO/UNESCO Legend for the Soil Map of the World as international standard. The WRB embraces all modern soil classification concepts, including Soil Taxonomy, the Legend of the FAO Soil Map of the World, the French Pedological Repository and the Russian concepts. The diagnostic criteria mostly match those of existing systems, so that correlation with national and previous international systems is plausible. The distribution and sample pictures of the Reference Soil Groups are based on work carried out by FAO, International Soil Reference and Information Centre (ISRIC) World Soils and the Universities of Leuven and Wageningen University.

9.2.3. USDA Soil Taxonomy

United States Department of Agriculture (USDA) Soil Conservation Service Soil Survey Staff led by Guy D. Smith started on a comprehensive system of soil classification in 1951. Several rounds of corrections (called, approximations) were made and in 1960, a comprehensive system, popularly called "7th Approximation" was published. In 1975, "Soil Taxonomy" was brought out by US Soil Survey Staff. It focused on soil properties that mostly can be measured quantitatively. The general objectives of this system were to aid easier remembrance of the characteristics of various soils, to understand the inter-soil relationships and to create a basis for prediction of soil behavior founded on principles of soil genesis.

9.3. Overview of USDA Soil Taxonomy

9.3.1. Categories

In Soil Taxonomy, there are six categories (Fig. 9.1) which, from top to bottom are, order, suborder, great group, subgroup, family and series. The highest categories have the fewest classes and criteria separating classes, in contrast to the lowest categories that have the most classes and criteria.

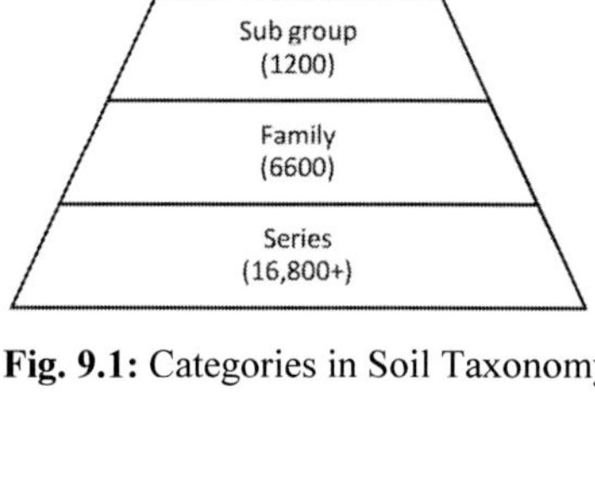

Fig. 9.1: Categories in Soil Taxonomy

Twelve classes are in the order level. The 12 orders and their characteristics are furnished in Table 9.1. The criteria used to differentiate orders are based on morphology as indicated by presence or absence of diagnostic horizons.

A **suborder** category is a subdivision of an order which is based on *genetic homogeneity*. Soil characteristics that distinguish suborders within an order vary with each order. For instance, soil moisture and temperature are the important factors that differentiate the suborders in the order Alfisol. The presence or absence of an argillic horizon, on the other hand, distinguishes the two suborders of the order Aridisol.

Table 9.1: Important properties of the orders

Order	Important Properties
Alfisol	Mineral soils relatively low in organic matter with relatively high base saturation. Contains horizon of illuvial clay. Moisture is available to mature a crop.
Aridisol	Mineral soils relatively low in organic matter. Contain developed soil horizons. Moisture is inadequate to mature a crop without irrigation in most years.
Entisol	Mineral soils lacking developed soil horizons. Moisture content varies.
Histosol	Soils composed mostly of organic matter. Moisture content varies.
Inceptisol	Mineral soils containing some developed horizons other than one of illuvial clay. Moisture is available to mature a crop.
Mollisol	Mineral soils with thick, dark surface horizons relatively high in organic matter and with high base saturation.
Oxisol	Mineral soils with no weatherable minerals. High in iron and aluminum oxides. Contain no illuvial horizons.
Spodosol	Soils that contain an illuvial horizon of amorphous aluminum and organic matter, with or without amorphous iron. Usually moist or well leached.
Ultisol	Mineral soils with an illuvial clay horizon. Has low base saturation. Generally found in humid climates.
Vertisol	Clayey soils with deep wide cracks at some time in most years. Moisture content varies.
Andisol	Volcanic ash soils
Gelisol	Frozen Tundra soils

A **great group** category is a subdivision of a suborder. Great groups are distinguished from one another based on *soil moisture, temperature regimes and diagnostic subsurface horizons*. For instance, all soils belonging to one of the suborders of Aridisols have argillic horizons. They might also have additional diagnostic horizons such as a petrocalcic. Soils having such additional horizons are placed in separate great groups.

Great group category is divided into three **subgroups**: *typic*, *intergrade* and *extragrade*. A typic subgroup truly represents the basic concept of the great group from which it derives. An intergrade subgroup encompasses soils of one great group that have some properties unique to soils in another great group or class. But these properties are not well developed or expressed to entail the inclusion of the soils in the great group toward which they grade. Extragrade subgroup soils have anomalous properties that do not intergrade to any known soil.

A soil **family** category is a group of soils within a subgroup that has analogous *physical and chemical properties* that influence response to management and manipulation. The principal characteristics used to differentiate soil families are texture, mineralogy and temperature, depth and calcareousness. Generally, family textural classes distinguish between clayey, loamy and sandy soils. For some soils the criteria also mention the amount of silt, sand and coarse fragments such as gravel, and cobbles.

Soil series category is a collection of soil individuals essentially uniform in differentiating characters such as color, texture, structure, consistency, pH and electrical conductivity and arrangement of horizons.

9.3.2. Nomenclature

One of the striking features of Soil Taxonomy is its system of nomenclature. Each class name is mnemonic, that is, to help memory and that would connote some properties of the soils of that class. The nomenclature is such that a person can recognize both the category of the class and the classes of the higher categories to which it belongs.

Class names were coined mostly from Greek and Latin roots that are familiar because of their use in many common words. Order names end in "sols". A formative element from each order (Table 9.2) is the ending of names of the suborders, great groups and subgroups.

Names of suborders have two syllables. The last syllable is the formative element from the order. The first syllable is the suborder formative element that suggests certain characteristics about the soil. A few examples and meanings of suborder formative elements are in Table 9.3.

Great group names are formed simply by including a great group formative element in front of the suborder name. The prefix is connotative and suggests the unique features of the particular great group named. For example, the great group name Durargids is derived from Aridisols with argillic horizons and duripans (dur + Argid = Durargids).

Table 9.2: Formative elements in orders

Order	Formative Elements	Derived from	Soils belonging to
Entisol	Ent	Recent	Azonal soils, Gley soils
Vertisol	Vert	L. *Verto*, turn	Grumusols, Black soils
Inceptisol	Ept	Inception	Brown forest soils
Aridisol	Id	L. *Aridus*, dry	Desert soils, Solonchalk, Solonetz
Mollisol	Oll	L. *Mollis*, soft	Chestnut, Chernozem soils
Spodosol	Od	Gk *Spodos*, Ash	Podsols
Alfisol	Alf	Non-sense syllable	Red soil, Planosol, Half-bog
Ultisol	Ult	L. *Ultimus*, last	Lateritic soils
Oxisol	Ox	Oxide soil	Laterite soils
Hisosol	Hist	Tissue soil	Peat soils. Bog soils
Andisol	And	Jap. *Ando*, black soil	Volcanic ash soils, dominated by allophane
Gelisol	El	Gk *Gelid*, Very cold	Frozen Tundra soils, permafrost

Names of subgroups are binomial that has an adjective before the name of the great group to indicate to which the subgroup belongs. "Typic" is added to a great group name to constitute the subgroup name for the soils that are typical of the great group. Thus, *Typic Durargids* is the name of the typical subgroup of the Durargids great group. An intergrade subgroup also carries the name of its great group, modified by the adjectival form of the name of the class towards which it grades. The adjective is framed by adding "ic" to the class name. For instance, Durargids that have some properties of the Haploxerolls great group are named as Haploxerollic Durargid at subgroup level. Extragrade subgroup names are framed by adding to the great group name a formative element that describes the extragradational feature. Abruptic Durargids, thus, include soils that have properties of Durargids but have, in addition, an abrupt transition between the A and B horizons characterized by a huge difference in clay percentage.

Family names are generally common terms that denote in more detail the features of a particular subgroup. Thus, a fine, montmorillonitic, thermic soil family is fine textured, with montmorillonite as the dominant clay mineral and has a thermic soil temperature regime.

Table 9.3: Formative elements in suborders

Formative Element	Derivative	Mnemonic	Connotation / Presence of
Alb	L. *Albus*, white	Albino	Albic horizon
Aqu	L. *Aqua*, water	Aquarium	Aquic moisture regime
Arg	L. *Argilla*, clay	Argillite	Argillic horizon
Bor	Gk *Boreas*	Boreal	Cool
El	Gk *Gel*, ice	Frost churning	OM; distorted boundary
Ferr	L. *Ferrum*, Fe	Ferruginous	Fe
Fibr	L. *Fibra*, fiber	Fibrous	Least decomposed stage
Fluv	L. *Fluvius*, river	Fluvial	Flood plain
Hem	Gk *Hemi*, half	Hemisphere	Inter-stage decomposition
Hum	L. *Humus*, earth	Humus	OM
Ochr	Gk. *pale*	Ocher	Ochric epipedon
Plag	*Plaggen*, sod	--	Plaggen epipedon
Psamm	*Psammos*, sand	Psammite	Sand texture
Rend	Rendzina	---	High carbonate content
Sapr	Gk *Sapros*, rotten	Saprophyte	Most decomposed stage
Torr	L. *Torridus*, dry	Torrid	Torric moisture regime
Ud	L. *Udus*, humid	---	Udic moisture regime
Umbr	L. *Umbra*, shade	Umbrella	Umbric epipedon
Ust	L. *Ustus*, burnt	Combustion	Ustic moisture regime
Xer	Gk *Xeros*, dry	Xerophyte	Xeric moisture regime

Example of the nomenclature of a soil series: Pilamedu (Plm) soil series in Tamil Nadu is referred to as: Pilamedu, fine, montmorillonitic isohyperthermic family of Typic Chromustert

Order	: ert	(Vertisol)
Suborder	: ust	(Ustic moisture regime)
Great group	: Chrom	(Colour value, moist, of 4 or more; colour value, dry, of 6 or more)
Subgroup	: Typic	
Family	: fine, montmorillonitic isohyperthermic family	
Series	: Pilamedu	

9.3.3. Diagnostic Horizons

Presence or absence of certain diagnostic horizons, which can be attributed to conditions of soil formation, is an important feature of Soil Taxonomy. These are horizons formed as a result of pedogenic process and have distinct properties. They are useful in identifying soils and classifying them in various categories, especially great groups. There are two types of diagnostic horizons, *viz., epipedons* and *endopedons.*

Epipedons are diagnostic horizons that have formed at the surface and are defined mostly in terms of soil color, content of organic matter and base saturation. They may be thinner than 'A' horizon, or may include 'E' horizon, or part or all of 'B' horizon. Mollic, anthropic, umbric, histic, ochric and plaggen are important diagnostic epipedons; mollic, ochric and umbric are important for soils of India.

Mollic Epipedon

- L. *Mollis* = soft
- It's dark, grassland soil
- Mineral horizon with $\geq$ 50% base saturation
- Has > 1% OM
- Color values darker than 5.5 when dry and 3.5 when moist
- Remains moist for > 9 months in a year

Anthropic Epipedon

- Gk *Anthropos* = Man
- Found in fields cultivated for long time with addition of OM
- Has > 250 ppm of citric acid soluble P_2O_5
- With or without 50% base saturation
- Remains moist for 3 months or more

Umbric Epipedon

- L. *Umbra* = shade / dark
- Mollic-like in thickness, organic matter content, colour, P_2O_5 content, consistence and structure
- But has < 50% base saturation
- Not naturally dry for > 3 months in a year

Histic Epipedon

- Gk *Histos* = Tissue
- Organic horizon
- Water saturated for long resulting in reduced conditions, unless artificially drained

- It's 20 to 60 cm thick
- Has low bulk density, often < 1Mg m^{-3}

Ochric Epipedon

- Gk *Ochros* = Pale, light coloured
- Light in colour
- Has < 1% OM
- It's dry for > 3 months in a year
- Ochric epipedon extends to first illuvial (B) horizon

Plaggen Epipedon

- Gk *Plaggen* = Sod
- Man-made horizon
- 50+ cm thick
- Resulted from centuries of accumulation of sod, straw and manure
- Commonly contains artifacts such as pottery and bricks

Endopedons

Argillic Horizon

- L. *Argilla* = Clay
- Horizon formed by illuvial clay
- At least one-tenth thick of all overlying horizons
- If overlying horizon has:

  <15% clay, argillic horizon has 3% more clay

  <15 to 40% clay, arg. horizon has 1.2x clay

  >40 % clay, arg. horizon has 8% more clay

Natric Horizon

- L. *Natrium* = Sodium
- High Na clay enriched
- Similar to argillic, but has sodium ≥15%, Sodium adsorption ratio ≥13

Agric Horizon

- L. *Ager* = Field
- Formed under a plough layer due to long & continued cultivation
- Clay and humus mixed
- Has lamellae (finger-shaped concentration of materials) of illuvial humus, clay and silt

Spodic Horizon

- Gk *Spodos* = Wood ash
- Humus and /or R_2O_3 enriched horizon with or without Fe
- Formed in cold humid regions

Oxic Horizon

- Horizon enriched with R_2O_3
- Dominance of 1:1 clay minerals from which Si has been leached
- At least 30cm thick
- Sandy loam or finer in texture
- <10% weatherable minerals

Cambic Horizon

- There should be some evidence to show some alteration of various properties (especially, colour) is on the way
- Illuviation not a dominant process
- Weatherable minerals present to the extent of 3 – 6%
- At least 10” below the surface soil, the cambic horizon should start

Sombric Horizon: Acidic, humus accumulation, tropical

Kandic Horizon: Argillic, with kaolinite-like clay

Albic Horizon

- Clay and free iron oxides have been removed
- Colour of the horizon is determined by the colour of the sand and silt particles
- Normally overlain by humus-enriched surface horizons (*mollic, umbric* or *ochric* horizons)

Sulphuric Horizon: High sulphides

Salic Horizon: High salt

Calcic Horizon: High $CaCO_3$

Gypsic Horizon: High $CaSO_4$

Petrocalcic Horizon

- A strongly cemented calcic horizon, with harness >3 on Moh's scale;
- Cemented by $CaCO_3$, not slaked in H_2O
- Not penetrable by spade or auger

Petrogypsic Horizon: A strongly cemented gypsic horizon, not slaked in H_2O

Placic Horizon:

- A thin (2 – 10 mm thick), lies within 50 cm of surface
- Dark reddish brown to black coloured, cemented by Fe and Mn

Duripan Horizon

- Cemented by Si in >50% of its volume
- Dissolves in concentrated basic solutions

Fragipan Horizon

- Brittle, located below an eluvial horizon
- Hard, very hard consistence when dry
- Mottled, slowly permeable to water

Plinthite Horizon: Fe-rich, humus-poor mixture of clay with quartz; dark red mottles

9.3.4. Soil Temperature and Moisture Regimes

The following are the soil temperature and moisture regimes used for placing soils at great group level.

Soil Temperature Regimes

- **Pergelic:** Mean annual soil temperature is <0 °C
- **Cryic:** Mean annual soil temperature is 0-8 °C
- **Frigid:** Mean annual soil temperature is 0-8 °C, but warmer in summer than cryic regime and the difference between mean summer and mean winter soil temperature is > 6 °C

- **Mesic:** Mean annual soil temperature is >8 - <15 ºC
- **Thermic:** Mean annual soil temperature is >15 - <22 ºC
- **Hyperthermic:** Mean annual soil temperature is >22 ºC

Soil Moisture Regimes

- **Aquic:** Soils water-saturated enough so that reducing conditions exist
- **Aridic or torric:** Soils dry >50% of time when not frozen; never moist >90 consecutive days when soil temperature is 8 ºC at 50cm depth
- **Perudic:** In most years these soils are not dry as long as 90 cumulative days
- **Udic:** Soils not dry as long as 90 days in a year
- **Ustic:** Soils dry for >90 days, but <180 days in a year
- **Xeric:** These soils have dry summers for >45 consecutive days and moist winters for >45 consecutive days in a year

10

Soil Texture

Physical properties of a soil greatly influence plant growth. The plant support, root penetration, drainage, aeration, retention of moisture and availability of plant nutrients are all linked with the physical condition of the soil. The physical properties of a soil which depend on the amount, size, shape, arrangement and mineral composition of its particles, also influence its chemical and biological behaviour. Important physical properties of soils are:

- Soil texture
- Soil structure
- Soil consistence
- Surface area
- Soil density
- Soil porosity
- Soil colour

10.1. Soil Texture

Soil texture refers to the relative proportion of soil particles. It is the relative per centage by weight of the three soil separates *viz.,* sand, silt and clay. In other words, it refers to the size of soil particles. The texture of a soil cannot be easily altered and it is considered as a basic property of a soil.

The soil separates are defined in terms of diameter in millimetres of the particles. Soil particles more than 2 mm in diameter are excluded from soil textural determinations. Stones and gravels may influence the way a land is used and managed because of the difficulties in tillage, but these larger particles make practically little or no contribution to the soil properties such as water holding capacity and capacity to store and supply plant nutrients.

Particles less than 2 mm are called *fine earth.* The components of fine earth are sand, silt and clay. The size limits of these fractions have been established by various organizations.

(a) United States Department of Agriculture

Soil separates diameter (mm)

Clay < 0.002 mm

Silt 0.002 – 0.05

Very fine sand 0.05 – 0.10

Fine sand 0.10 – 0.25

Medium sand 0.25 - 0.50

Coarse sand 0.50 - 1.00

Very coarse sand 1.00 – 2.00

(b) International Union of Soil Sciences

Soil separates diameter (mm)

Clay < 0.002 mm

Silt 0.002 – 0.02 mm

Fine sand 0.02 – 0.2 mm

Coarse sand 0.2 – 2.0 mm

10.1.1. Sand

Composition-wise, sand generally consists of quartz, but may also contain fragments of feldspar, mica and, occasionally, heavy minerals such as zircon, tourmaline and hornblende. Sand has uniform dimensions and, therefore, can be considered as spherical; but not necessarily smooth and has jagged surface. Sand feels "gritty". Spaces between sand particles allow free drainage and air entry. Sand particles thus hold little water and therefore are prone to drought. They hold fewer nutrients than do silt or clay.

10.1.2. Silt

Silt has particle size intermediate between sand and clay; feels soft, silky or soapy and is approximately spherical or cubical in shape; since the size is smaller, the total surface area is more; has the physico-chemical properties as that of clay to a limited extent. Silt holds more nutrients, retains more water than sand.

10.1.3. Clay

Soil particles that are less than 0.002 mm in diameter are clay particles. They are plate-like or needle-like in shape and belong to alumino-silicate group of minerals. Clays are secondary minerals derived from primary minerals in the rock. Though clays have high adsorption of water it is not available readily to crops. Very small pore spaces make movement of water and air slow.

10.1.4. Loam

An ideal loam contains 40% sand, 40% silt and 20% clay and is the most productive soil for farming crops. Loam has good water-holding capacity (from clay), good drainage (from sand) and can be nutritious (from silt).

10.2. Stokes' Law

Stokes (1851) proposed an equation that expresses the settling velocities of small spherical particles in a fluid medium. The law is derived by consideration of the forces acting on a particular particle as it sinks through a liquid column under the influence of gravity. The magnitude of the force resisting the fall is equal to $6\pi r\eta v$, in which r is the radius of the sphere, η is the viscosity of the liquid, and v is the velocity of fall. The force acting downward is equal to $\frac{4}{3}\pi r^3 (d - d')g$, in which d is the density of the sphere, d' is the density of the liquid, and g is the gravitational constant. At a constant velocity of fall, the upward and downward forces balance each other. Equating the two expressions given above and solving for v therefore yields the velocity, expressed by Stokes's law as $v = {}^2/_9(d - d')\, g\, r^2/\eta$ to suggest the relationship between the radius of the particles and its rate of fall in a liquid: "The velocity of a falling particle in a liquid is proportional to the square of its radius".

Assumptions and Limitations of Stokes' Law:

1. Particles are rigid and spherical / smooth. This requirement is very difficult to fulfill, because the particles are not completely smooth over the surface and spherical.
2. The particles are large in comparison with the molecules of the liquid so that in comparison with the particle the medium can be considered as homogenous. The particles must be big enough to avoid Brownian movement. Particles less than 0.0002 mm exhibit Brownian movement so that the rate of falling is varied.
3. The fall of the particles is not hindered or affected by the proximity of the wall of the vessel or of the adjacent particles. Many fast falling particles may drag finer particles down along with them.
4. Particles must be of uniform density. But this is usually not so because of their different chemical and mineralogical composition.
5. The suspension must be still. Any movement in the suspension will alter the velocity of fall and such movement is brought by the sedimentation

of larger particles (> 0.014 mm). They settle so fast and create turbulence in the medium.

10.3. Methods of Determination of Texture

Numerous methods for laboratory and field use have been developed. Analytical procedure by which the soil particles are separated is called, "mechanical analysis".

- Elutriation method – Water & air
- Decantation/ beaker method
- The hydrometer method
- Feel method
- International pipette method [*Robinson, G. W. 1922. A new method for the mechanical analysis of soils and other dispersions. J. Agr. Sci. 12: 306-321.*]

Feel Method

Evaluated by attempting to squeeze the moistened soil into a thin ribbon as it is pressed with rolling motion between thumb and pre finger or alternately to roll the soil into a thin wire. Four aspects to be determined: (i) feel by fingers, (ii) ball formation, (iii) stickiness and (iv) ribbon formation (Fig. 10.1).

10.4. Soil Textural Classes

In order to convey the textural make up of soils and to give an indication of their physical properties, soil textural class names are used. According to the relative proportion of the three main fractions (sand, silt and clay) present, a soil is given a name to indicate its texture. Such a name not only gives an idea of the textural composition of a soil but also of its various properties in general. On this basis soils are classified into various textural classes like sands clays, silts, loams, etc.

10.4.1. Sand

The sand group includes all soils in which the sand separates make up at least 70% and the clay separates 15% or less of the material by weight. The properties of such soils are therefore characteristically those of sand in contrast to the stickier nature of clays. Two specific textural classes recognized in this group are sandy loam and loamy sand.

10.4.2. Silt

This group includes soils with at least 40% silt and 12% or less clay. Naturally those of silt dominate the properties of this group. Silt is the only one textural class in this group.

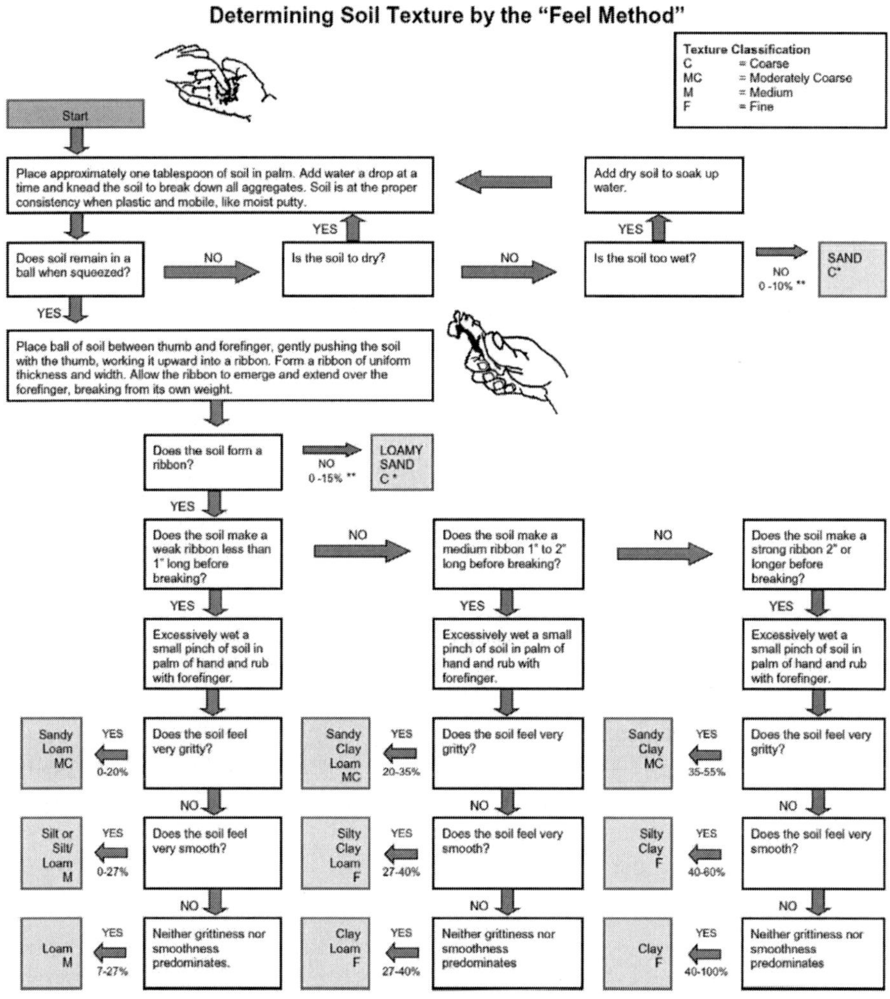

* Sand Particle size should be estimated (very fine, fine, medium, coarse) for these textures. Individual grains of very fine sand are not visible without magnification and there is a gritty feeling to a very small sample ground between the teeth. Some fine sand particles may be just visible. Medium sand particles are easily visible. Examples of sand size descriptions where one size is predominant are; very fine sand, fine sandy loam, loamy coarse sand.

** Cay percentage range.

Fig.10.1: Determination of soil texture by feel method

10.4.3. Clay

Clay is the smallest mineral particle in soil. On equal weight basis, clay particles have about 1,000 times as much external surface area as sand. Sand and silt particles are much bigger than clay in size and are quite inactive chemically because of their mineral composition and limited surface area. Clay is the storehouse of plant nutrients owing to its large surface area and mineral composition. Therefore, soils with more clay have more nutrients than sandy soils. Clay particles are the active fraction of a soil, because chemical reactions occur at their surface. The chemical reactions control the adsorption and release of plant nutrients and many other chemicals in the environment. To be designated as clay and in order that the characteristics of the clay separates are distinctly dominant, a soil must contain at least 35% of the clay separate and, in most cases, not less than 40%. The class names are clay, sandy clay and silty clay. Sandy clays may contain more sand than clay. Likewise, the silt content of silty clays usually exceeds clay fraction.

10.4.4. Loam

An ideal loam may be defined as a mixture of sand (40%), silt (40%) and day (20%) particles that exhibits the properties of those separates in about equal proportions. Loam soils do not exhibit dominant physical properties of sand, silt or clay. Loam does not contain equal percentage of sand, silt and clay; however, may exhibit approximately equal properties of sand, silt and clay.

10.5. Determination of Textural Class

As per the system developed by the United States Department of Agriculture, there are twelve textural classes. The textural triangle (Fig. 10.2) is used to determine the soil textural name after the per centages of sand, silt and clay are determined from a laboratory analysis. Since the soil's textural classification includes only mineral particles and those of less than 2 mm diameter, the sand plus silt plus clay per centages equal 100 per cent. (Note: Organic matter is not included.) Knowing the amount of any two fractions automatically fixes the per centage of the third one.

To use the diagram, locate the percentage of clay first and project inward parallel to sand line. Do likewise for silt and project inward parallel to clay line and for sand, project inward parallel to silt. The point at which the projections cross or intersect will identify the class name.

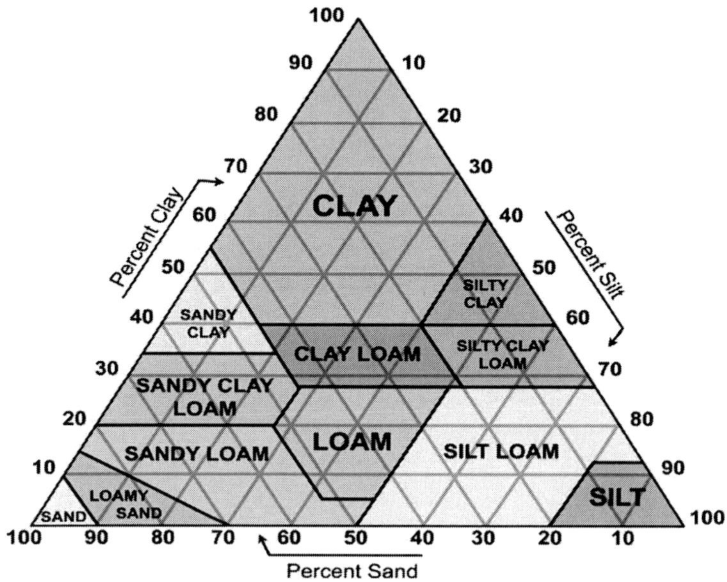

Fig. 10.2: Textural triangle

Assume that you have a soil that is 35 per cent clay, 52 per cent silt and 13 per cent sand. The per cent of clay is identified on the left side of the triangle. From the lower left corner to the top of the triangle, the per cent clay increases from 0 to 100 per cent. Move along the left side of the triangle until you reach 35 per cent clay. Then draw a line at 35 per cent clay that is parallel to the bottom of the triangle. The per cent silt is identified along the right side of the triangle. From the top of the triangle to the lower right, the per cent silt increases from 0 per cent to 100 per cent. Move along the right side of the triangle until you reach 52 per cent silt. Now draw a line at 52 per cent silt that is parallel to the left side of the triangle. The bottom of the triangle identifies the per cent sand. From the lower right corner to the lower left corner, the per cent sand increases from 0 per cent to 100 per cent. Move along the bottom of the triangle until you reach 13 per cent sand. Draw a line at 13 per cent sand that is parallel to the right side of the triangle. The point at which these three lines intersect will define the soil's texture, which, in this case is *silty clay loam*. Sometimes, the intersecting point may exactly fall on the line between the textural classes. Then it is customary to use the name of the finer fraction when it happens. For example, a soil containing 40% clay, 30% sand and 30% silt is designated as clay rather than clay loam.

10.6. Importance of Soil Texture

- Texture has good effect on the productivity of soil.
- Sandy soils are of open character, usually loose and friable. Such type of texture is easy to handle in tillage operations. Sand facilitates drainage and aeration. It allows rapid evaporation and percolation. Sandy soils have very little water holding capacity. Such soils cannot stand drought and are unsuitable for dry farming. Sandy soils are poor storehouse of plant nutrients and contain low organic matter. Leaching of applied nutrients is very high. In sandy soil, few crops can be grown such as potato, groundnut and cucumbers.
- Clayey soils are difficult to till and require much skill in handling. When moist, clayey soils are exceedingly sticky and when dry, become very hard and difficult to break. They have fine pores, and are poor in drainage and aeration. They have a high water holding capacity and poor percolation, which usually results in water logging. They are generally very fertile soils, in respect of plant nutrient content. Rice, jute, sugarcane can be grown very successfully in these soils.
- Loam and silt loam soils are highly desirable for cultivation.
- Generally, the best agriculture soils are those that contain 10 – 20 % clay, 5 – 10 % organic matter and the rest equally shared by silt and sand.

11

Soil Structure

Soil Structure Defined

Soil structure is defined as the arrangement of individual soil particles with respect to each other into a pattern. Structure describes the gross, over-all arrangement of the soil solids. In other words, soil structure is the 'architecture' of a soil.

Soils are composed of structural units, called "*peds*". The peds consist of many individual particles, bound together by cementing agents like organic matter, clay and hydrous oxides of iron and aluminium. However, the strength of the bonds, the size and shape of the peds and the proportion of the soil particles involved in the peds differ considerably among soils.

11.1. Types of Structure

1. *Structureless or Simple structure*

Natural cleavage planes absent or indistinct

a. Single grain structure (only in sand and silt)

b. Massive structure (similar to single grain, but coherent) [soil crusts, fragipans]

2. *Structured or Compound structure*

Natural cleavage planes distinct. Description of peds is made from relative lengths of vertical and horizontal axes and by contour of edges.

There are four principal types of compound structure:

Plate-like (Platy)

In this type, the aggregates are arranged in relatively thin horizontal plates or leaflets. The horizontal axis is longer than the vertical axis. When the units/layers are thick they are called "*platy*". When they are thin then it is "*laminar*". Platy structure is mostly found in the surface layers of virgin soils but may be present in the subsoil and is inherited from the parent material, especially by the action of water or ice.

Prism-like

The vertical axis is more developed than the horizontal axis, resulting in a pillar like shape. Varying in length from 1- 10 cm, they commonly occur in sub soil horizons of arid and semi-arid regions. When the tops are rounded, the structure is termed as "columnar"; when the tops are flat / plane, level and clear-cut, the structure is "prismatic".

Block- like

All three dimensions are about the same size and irregularly six faced. When the faces are flat and corners and edges are sharp and distinct, the structure is named as "angular blocky". When the faces and corners are rounded and edges are sharp, it is called "subangular blocky". These types usually are confined to the subsoil and characteristics have much to do with soil drainage, aeration and root penetration.

The structure is "granular" when both the corners and edges are rounded. The sharp edges of peds are due to physical forces [wetting, drying, freezing, thawing, etc.]. The rounded surfaces of peds are due to organic matter.

When the granules are very porous, it is termed as crumb. This is specific particularly to surface soil with high organic matter and grassland soils.

11.2. Classes of Structure

Each primary structural type of soil is differentiated into 5 size classes depending upon the size of the individual peds. The terms commonly used for the size classes are:

- Very fine or very thin
- Fine or thin
- Medium
- Coarse or thick
- Very Coarse or very thick

For platy types, the terms thin and thick are used, while for other structural types the terms fine and coarse are used.

11.3. Grades of Structure

Grades denote the degree of distinctness of the individual peds. It is determined by the stability of the aggregates. Grade depends on moisture content of the soil, organic matter, texture, etc. There are four terms that are commonly used to describe the grade of soil structure:

- *Structureless:* There is no noticeable aggregation, such as conditions exhibited by loose sand.
- *Weak structure:* Poorly formed, indistinct formation of peds, which are not durable and much unaggregated material.
- *Moderate structure:* Moderately well-developed peds, which are fairly durable and distinct.
- *Strong structure:* Very well formed peds, which are quite durable and distinct.

In naming a soil structure the sequence is followed: grade, class and type; for example: strong coarse angular blocky, moderate thin platy, weak fine prismatic.

Table 11.1: Types, kind and classes of soil structure

Type	Kind	Class	Size (mm)
Structureless: No observable aggregation	Single-grain structure: Loose, incoherent mass of individual particles in sands		
	Massive structure: A coherent mass showing no evidence of any distinct arrangement of soil particles, hard to break part		
Block-like: Particles are arranged around a point and bound by flat or rounded surfaces	Blocky (angular blocky): Corners and edges are sharp and distinct	Fine blocky	< 10
		Medium blocky	10 - 20
		Coarse blocky	20 - 50
		Very coarse blocky	> 50
	Subangular blocky: Corners are rounded and edges are sharp	Fine subangular blocky	< 10
		Medium subangular blocky	10 - 20
		Coarse subangular blocky	20 - 50
		Very coarse s. a. blocky	> 50
	Granular: Both corners and edges are rounded	Fine granular	< 2
		Medium granular	2 - 5
		Coarse granular	> 5
Plate-like	Platy structure: Horizontal planes developed	Fine platy	< 2
		Medium platy	2 - 5
		Coarse platy	> 5
Prism-like	Prismatic structure: Vertical planes well developed; edges sharp	Fine prismatic	< 10
		Medium prismatic	10 - 20
		Coarse prismatic	20 - 50
		Very coarse prismatic	> 50
	Columnar structure: Vertical planes well developed; edges near top of the column not sharp	Fine columnar	< 10
		Medium columnar	10 - 20
		Coarse columnar	20 - 50
		Very coarse columnar	> 50

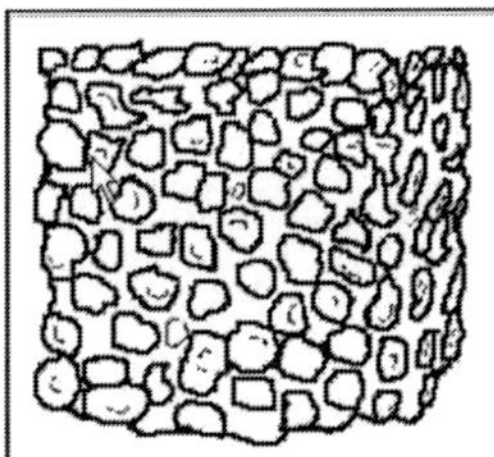	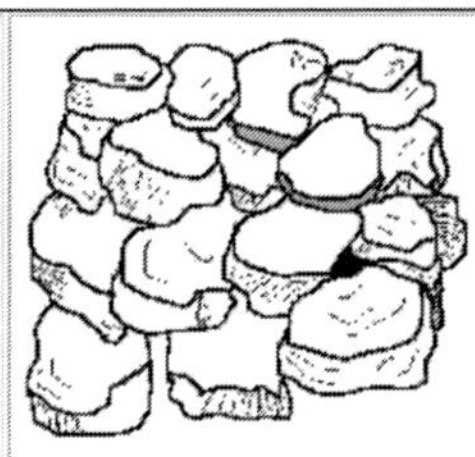	
Granular: Resembles cookie crumbs and is usually less than 0.5 cm in diameter. Commonly found in surface horizons where roots have been growing.	**Blocky**: Irregular blocks that are usually 1.5 - 5.0 cm in diameter.	**Prismatic**: Vertical columns of soil that might be a number of cm long. Usually found in lower horizons.
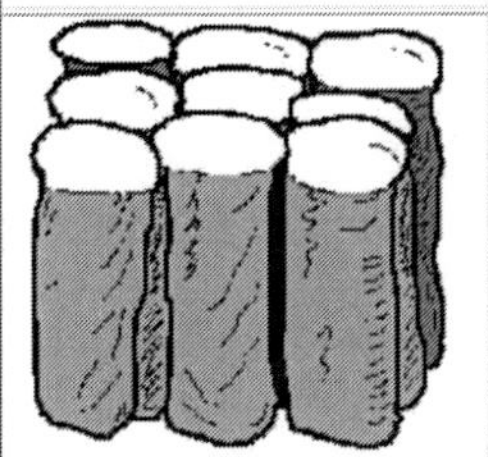		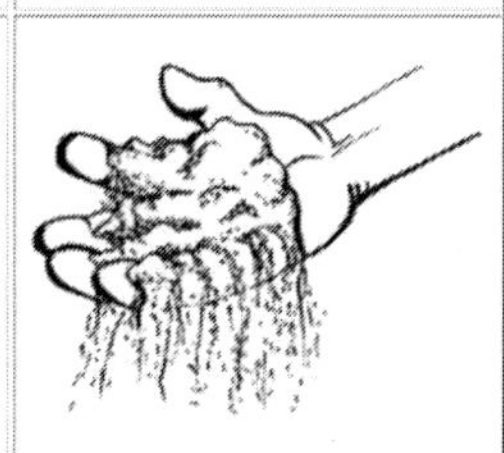
Columnar: Vertical columns of soil that have a salt "cap" at the top. Found in soils of arid climates.	**Platy**: Thin, flat plates of soil that lie horizontally. Usually found in compacted soil.	**Single Grained**: Soil is broken into individual particles that do not stick together. Always accompanies a loose consistence. Commonly found in sandy soils.

Fig. 11.1: Types and kind of soil structure

11.4. Genesis of Soil Structure

The genesis of soil structure starts with the bonding of the soil particles into structural unit. The bonding between individual particles in the structural units is generally stronger than the structural units themselves.

During aggregate formation, a number of primary particles such as sand, silt and clay are brought together by the cementing or binding effect of soil colloids. Colloidal clay, iron and aluminium hydroxides and decomposing organic matter are the cementing materials taking part in aggregate formation. Irrespective of the nature of the cementing material, it is ultimately the dehydration of colloidal matter accompanied with pressure that completes the process of aggregation.

Clay particles play a key role in the formation of soil aggregates by virtue of their high surface area and surface charge. Sand and silt particles cannot form aggregates as they do not possess the power of adhesion and cohesion. These particles usually carry a coating of clay particles and are enmeshed in the aggregates formed by the adhering clay particles. Colloidal particles form aggregates only when they are flocculated. There is vast difference between flocculation and aggregation.

Flocculation is brought about by coalescence of colloidal particles and is the first step in aggregation. Aggregation is something more than flocculation involving a combination of different factors such as hydration, pressure, dehydration etc., and required cementation of flocculated particles. The cementation is caused by cations, oxides of Fe and Al, humus substances and products of microbial excretion and synthesis. Clay particles form aggregates only if a liquid like water, whose molecules possess an appreciable dipole moment, wets them.

Clay - – + Water - - + Cation + – -Clay - – + Water - - + Cation+ - -Clay -

Aggregation also depends upon the nature, size and amount of clay particles, dehydration of clay particles, cations like calcium and anions like phosphate. The colloidal Fe oxides and Al oxides act as cementing agent in aggregation and bind the sand and silt particles.

Organic matter plays significant role in forming soil aggregates. During decomposition cellulosic substances produce a sticky material very much akin to mucus or mucilage. The stickiness may be due to the presence of humic acid or related compounds produced. Besides, certain polysaccharides formed during decomposition also aid in soil aggregation. Some fungi and bacteria exhibit cementing effect, probably due to the presence of slimes and gums on the surface of the living organisms produced as a result of the microbial activity.

11.5. Factors Affecting Soil Structure

The development of structure in arable soil depends on the following factors:

1. *Climate*: Climate has significant influence on the degree of aggregation and on the type of structure. In arid regions there is very little aggregation of primary particles. In semiarid regions, the degree of aggregation is greater.
2. *Organic matter*: Organic matter improves the structure of both sandy soils and clay soils. In sandy soil, the sticky and slimy materials produced by the decomposing organic matter and the associated microorganisms cement the sand particles together to form aggregates. In clayey soil, organic matter modifies the properties of clay by reducing its cohesiveness, thus making clay crumbier.
3. *Tillage*: Tillage implements break large clods into smaller fragments and aggregates. For good granular and crumby structure, moisture content in the soil should be optimum. If the moisture content is too high, large

clods will be formed on drying; if it is too low, some of the existing aggregates will be broken down.

4. *Plants, Roots and Residues*: Excretion of gelatinous organic substances and exudates from roots serve as a link. Root hairs make soil particles to cling together. Pressure exerted by the roots also holds the particles together. Dehydration of soil strains the soil due to shrinkage, results in cracks and leads to aggregation. Plant residues shade the soil, preventing it from extreme and sudden temperature and moisture changes as well as from rain drop impedance. Plant residues also serve as a food to microbes which are the major aggregate builders.

5. *Animals*: Among the soil fauna small animals like earthworms, moles and insects etc. that burrow in the soil are the chief agents that take part in the aggregation of finer particles.

6. *Microbes*: Algae, fungi and actinomycetes keep the soil particles together. Fungi and actinomycetes exert mechanical binding by mycelia, while bacteria aid in cementation by their products of decomposition and synthesized materials.

7. *Wetting and drying*: When a dry soil is wetted, the soil colloids swell on absorbing water. Drying and shrinkage exert strains on the soil mass, giving rise to cracks which break the soil mass into clods and granules of various sizes.

8. *Exchangeable cations*: Exchangeable cations like Ca and Mg have flocculating effects, while Na^+ deflocculates.

9. *Inorganic cements*: $CaCO_3$ and sesquioxides.

11.6. Soil Strength

Soil strength is defined as *the amount of force required to move or rearrange the soil particles*. Soil strength represents the reaction of a soil to an applied force. It is an important physical property affecting seedling emergence and plant growth. High-strength soils resist deformation (compaction especially), break-up (shearing and shattering), and slippage. Strength to a soil is by virtue of:

- cohesive forces between particles; and
- frictional resistance met by particles that are forced to slide over one another, or move from interlocked positions.

The three main factors that influence soil strength are soil moisture, texture and bulk density. The lower the soil moisture content, or drier the soil, the greater is the soil strength. In soils that are saturated or water-logged, the soil strength is weak. Soil strength of aggregated soils increases with the increase in the clay content increases. Poorly aggregated soils or single-grained soils (sand, sandy loam and loamy sand) usually are weakest in soil strength, unless they are cemented or compacted. The most widely used device in the field to measure soil strength is cone penetrometer.

12

Soil Consistence

Soil Consistence Defined

Soil consistence is the strength with which soil materials are held together or the resistance of soils to deformation and rupture. Soil consistence is measured for wet, moist and dry soil samples. For wet soils, it is expressed in terms of both stickiness and plasticity. Stickiness is the ability of soil materials to adhere to other objects. Plasticity is the ability of the soil materials to change shape continuously under a constant pressure and to retain the impressed shape when the pressure is removed. Soil consistence may be estimated in the field using simple tests or may be measured more accurately in the laboratory.

12.1. Determination of Wet-Soil Consistence

Testing for wet-soil consistence is done when the soil is saturated with water, as, for example, immediately after a good rainfall. First, stickiness and then plasticity, is determined.

12.1.1. Field Test for Stickiness of Wet Soil

A small amount of wet soil is pressed between the thumb and forefinger to see if it sticks to fingers. Then slowly the fingers are opened. The stickiness is rated as follows:

0 Non-sticky, if no soil or practically no soil sticks to the fingers;

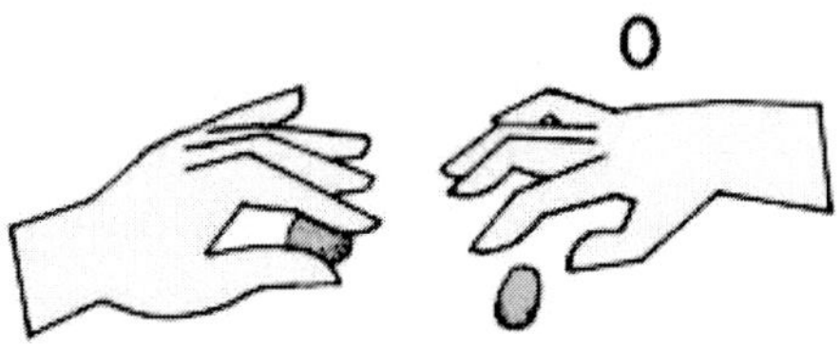

1 Slightly sticky, if the soil begins to stick to the fingers but comes off one or the other cleanly and does not stretch when the fingers are opened;

1

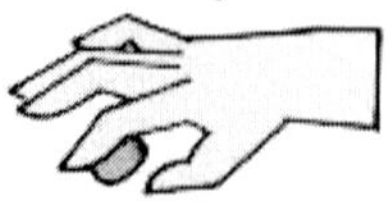

2 Sticky, if the soil sticks to both the thumb and forefinger and tends to stretch a little and pull apart rather than pulling free from the fingers;

2

3 Very sticky, if the soil sticks firmly to both thumb and forefinger and stretches when the fingers are opened.

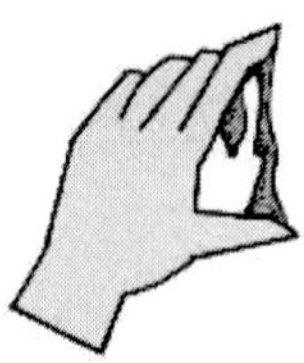

3

12.1.2. Field Test for Plasticity of Wet Soil

A small amount of wet soil is rolled between the palms of hands until it forms a long, round strip like a wire about 3 mm thick. The plasticity is rated as follows:

0 Non-plastic, if no wire can be formed;

1 Slightly plastic, if a wire can be formed but can easily be broken and returned to its former state;

2 Plastic, if a wire can be formed but, when it is broken and returned to its former state, it cannot be formed again;

3 Very sticky, if the soil sticks firmly to both thumb and forefinger and stretches when the fingers are opened.

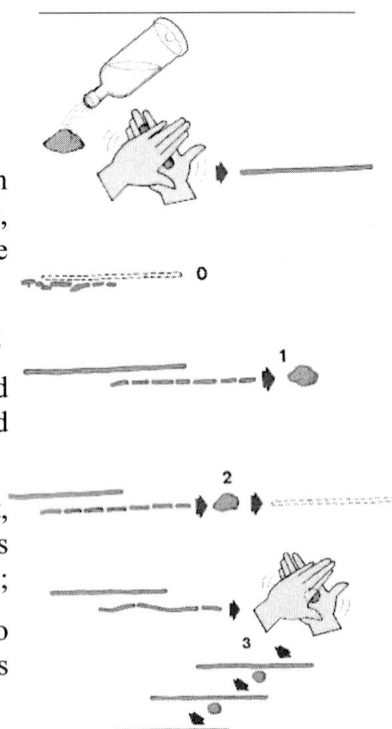

12.2. Determination of Moist-Soil Consistence

Field testing is done when the soil is moist but not wet, as, for example, 24 hours after a good rainfall. A small amount of moist soil is tried to be crushed by pressing it between the thumb and forefinger or by squeezing it in the palm of the hand. Moist soil consistence is rated as follows:

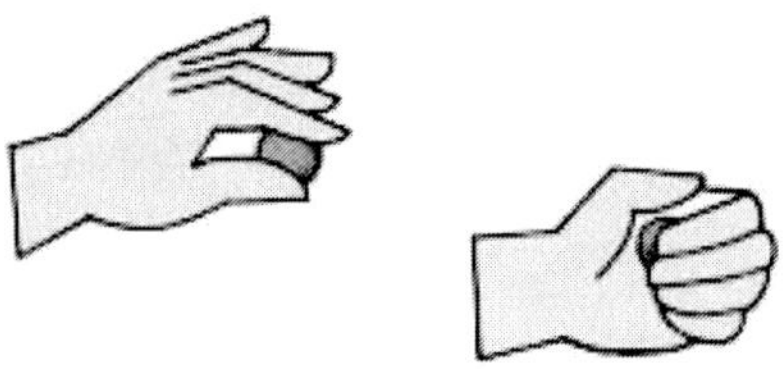

0 Loose, if the soil is non-coherent (single-grain structure);

1 Very friable, if the soil crushes easily under very gentle pressure but will stick together if pressed again;

2 Friable, if the soil crushes easily under gentle to moderate pressure;

3 Firm, if the soil crushes under moderate pressure but resistance is noticeable;.

4 Very firm, if the soil crushes under strong pressure, but this is difficult to do between the thumb and forefinger;

5 Extremely firm, if the soil crushes only under very strong pressure, cannot be crushed between the thumb and forefinger, but must be broken apart bit by bit.

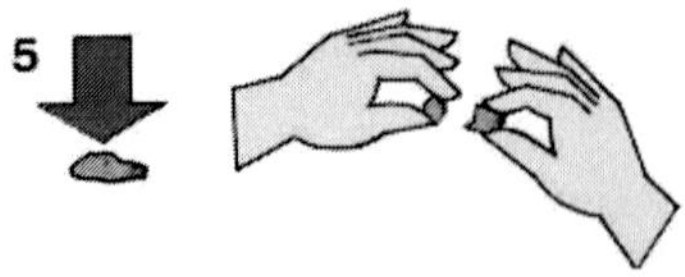

12.3. Determination of Dry-Soil Consistence

Field testing is done when the soil has been air-dried. A small amount of dry soil is tried to be broken by pressing it between the thumb and forefinger or by squeezing it in the palm of the hand. Dry soil consistence is rated as follows:

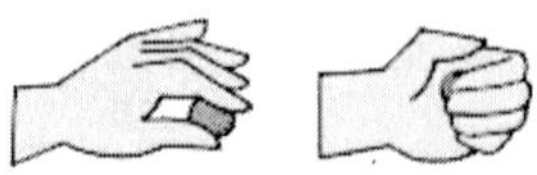

0 Loose, if the soil is non-coherent (single-grain structure):

1 Soft, if the soil is very weakly coherent and friable. breaking to powder or individual grains under very slight pressure;

2 Slightly hard, if the soil resists light pressure, but can be broken easily between thumb and forefinger;

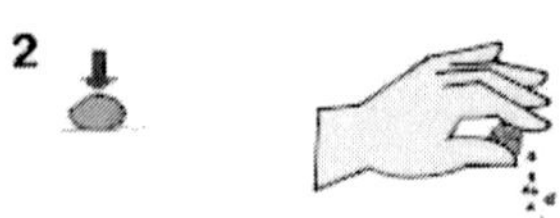

3 Hard, if the soil resists moderate pressure, can barely be broken between the thumb and forefinger, but can be broken in the hands without difficulty;

4 Very hard, if the soil resists great pressure, cannot be broken between the thumb and forefinger but can be broken in the hands with difficulty;

5 Extremely hard, if the soil resists extreme pressure and cannot be broken in the hands.

12.4. Laboratory Determination of Soil Consistence

The consistence of a soil changes with the amount of water present and such changes can be accurately measured in the laboratory adopting standard procedures that determine the Atterberg Limits. An Atterberg Limit corresponds to the moisture content at which a soil sample changes from one consistence to another. These limits may then be used for judging the suitability of the soil, such as for the construction of small earth-dams and pond dikes.

- Liquid consistence - fluid or liquid mud;
- Plastic consistence - kneading and moulding to shape are possible;
- Semi-solid consistence - no more kneading is possible and the volume decreases (shrinkage) as the sample dries up.

Atterberg Limits - Soil Consistence

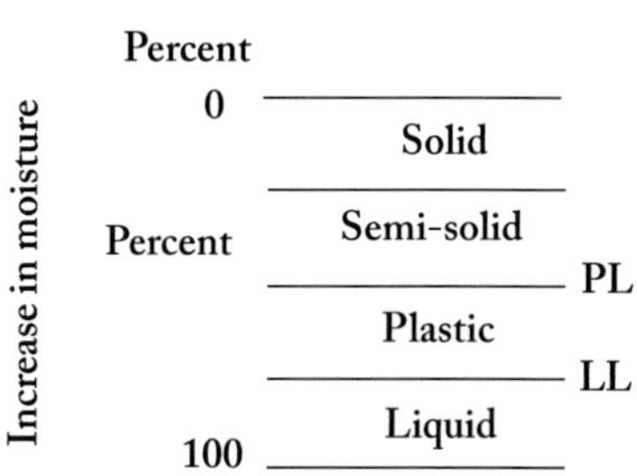

The Liquid Limit (LL)

The liquid limit is the per cent moisture content at which a soil changes with decreasing wetness from the liquid to the plastic consistence or with increasing wetness from the plastic to the liquid consistence.

The Plastic Limit (PL)

The plastic limit is the per cent moisture content at which a soil changes with decreasing wetness from the plastic to the semi-solid consistence or with increasing wetness from the semi-solid to the plastic consistence. The plastic limit denotes the lower limit of the plastic state. A small quantum increase in moisture above the plastic limit will destroy the cohesion of the soil.

The above limits can easily be determined in the laboratory, using disturbed or undisturbed samples. They provide very useful information for classifying soils. Both the liquid and plastic limits depend upon the amount and type of clay present in the soil:

- A soil with a high clay content usually has high LL and PL;
- Colloidal clays have higher LL and PL than non-colloidal clays;
- Sand, gravel and peat have no plasticity, their PL= 0;
- Silts have plasticity only occasionally, their PL being equal to or slightly greater than 0.

A typical laboratory test showing average LL and PL (per centage moisture) is given below:

Soil type	LL	PL
Sands	20	0
Silts	27	20
Clays	100	45
Colloidal clays	399	46

12.5. Field Determination of Plastic Limit - The Thread Method

A soil sample is taken and allowed to dry;

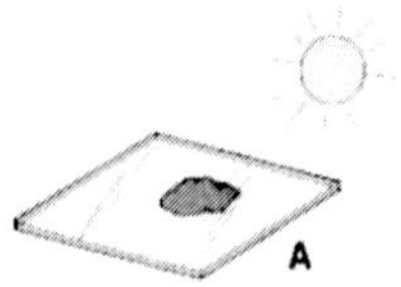

If you do not succeed, add a little more water and try again;

This process is repeated, adding a little more water each time, until a thread can be rolled. The water content will then correspond to the plastic limit and can be expressed as a per cent of the weight of the sample.

A little water is added to the soil sample and it is rolled on a flat surface like a small glass plate. Try to form a thread 3 mm thick and 10 cm long without breaking it;

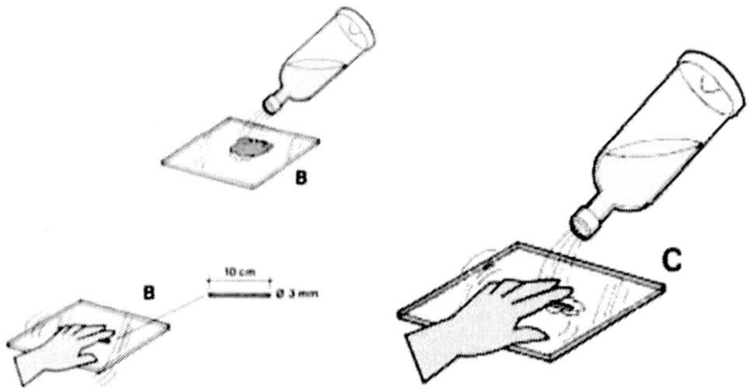

12.6. Plasticity Index and its Significance

On the basis of the liquid limit and the plastic limit, the **plasticity index** (PI) can be defined as the numerical difference between them:

$$PI = LL - PL$$

The plasticity index is expressed in per cent of the dry weight of the soil sample. It shows the size of the range of the moisture contents at which the soil remains plastic. The plasticity index generally depends *only* on the amount of clay present, indicating the fineness of the soil and its capacity to change shape without altering its volume. A high PI indicates an excess of clay colloids in the soil. Its value is zero whenever the PL is greater or equal to the LL. The plasticity index is also a good indicator of soil compressibility. The greater the PI, the greater is the soil compressibility. Examples of plasticity index values are given below for reference:

Soil	PI (%)	Degree of plasticity
Sand or silt	0 - 1	Non-plastic
▪ Traces of clay	1 - 5	Slight plasticity
▪ Little clay	5 - 10	Low plasticity
Clay loam	10 - 20	Medium plasticity
Silty clay	20 - 25	High plasticity
Clay	> 35	Very high plasticity

13

Soil Density and Porosity

Soil Density

The density of a soil, as any other matter, is its mass per unit volume. However, in case of soil, because of its unique nature, the density can be expressed by two terms, *viz.*, bulk density and particle density. Soil density is typically expressed in mega gram per cubic metre (Mg m^{-3}) which is numerically equal to g cm^{-3}.

13.1. Bulk Density

Soil bulk density is defined as the dry weight of soil per unit volume of soil, including pore space., Particle density, on the other hand, considers only the mineral solids, excluding pore space. Fig. 13.1 illustrates the difference between bulk density and particle density.

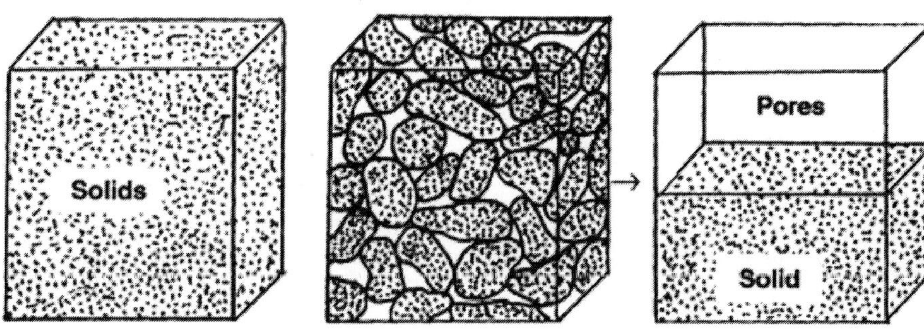

Particle Density
100% solid
Weight = 2.66 g
Volume = 1 cm^3

Bulk Density
50% solid, 50% pore space Weight = 1.33 g Volume = 1 cm^3

Fig. 13.1: Bulk density and particle density

One-half of an ideal soil is solids and the remaining half is pore space. Using an example of a 1 cm^3 volume, the ideal soil would have 0.5 cm^3 of pore space and 0.5 cm^3 of solids. Pore space filled with air weighs 0 g. Since organic matter is a very small portion of the solids, it is usually ignored in this calculation. The mineral solids would weigh 1.33 g when dry and is determined by multiplying particle density by the volume of solids: 2.66 g cm^{-3} x 0.5 cm^3 = 1.33 g. The bulk density, then, is the dry weight of soil divided by the volume of soil: 1.33 g ÷ 1 cm^3 = 1.33 g cm^{-3}.

Bulk density is an indicator of soil compaction. It is calculated as the dry weight of soil divided by its volume. This volume includes the volume of soil particles and the volume of pores among soil particles. Bulk density is typically expressed in mega gram per cubic metre (Mg m^{-3}) which is numerically equal to g cm^{-3}.

The importance of soil bulk density is that it reflects the soil's ability to function for structural support, water and solute movement, and soil aeration. Bulk densities above thresholds indicate impaired function (Table 13.1). Bulk density can be used to convert between weight and volume of soil. It is also useful to express soil physical, chemical and biological measurements on a volumetric basis for soil quality assessment and comparisons between management systems, thus ensuring the validity of comparisons by removing errors associated with differences in soil density at time of sampling.

Table 13.1: General relationship of soil bulk density to root growth

Soil Texture	Ideal bulk densities for plant growth (Mg m^{-3})	Bulk densities that restrict root growth (Mg m^{-3})
Sandy	< 1.60	> 1.80
Silty	< 1.40	> 1.65
Clayey	< 1.10	> 1.47

High bulk density indicates low soil porosity and soil compaction. It may restrict root growth, and movement of air and water through the soil. Compaction can lead to shallow plant rooting and poor plant growth, eventually reducing crop yield and reducing vegetative cover available to protect soil from erosion. By reducing water infiltration into the soil, compaction entails increased runoff and erosion from sloping land or waterlogged soils in flatter terrain. In general, some degree of soil compaction that restricts water movement through the soil profile is beneficial under arid conditions, but under humid conditions compaction decreases yields. The following practices can lead to poor bulk density:

- Consistently plowing or disking to the same depth.
- Allowing equipment traffic, especially on wet soil.
- Using a limited crop rotation without variability in root structure or rooting depth.
- Incorporating, burning, or removing crop residues.
- Overgrazing forage plants, and allowing development of livestock loafing areas and trails.

- Using heavy equipment for building site preparation or land smoothing and leveling.

Any practice that improves soil structure decreases bulk density; however, in some cases these improvements may only be temporary. For example, tillage at the beginning of the growing season loosens compacted soil layers and decreases bulk density; but, subsequent trips across the field by farm equipment, rainfall events, animals and other disturbances can recompact soil. On cropped land, long-term solutions to soil compaction problems hinge on decreasing soil disturbance and increasing soil organic matter. A management system that uses cover crops, crop residues, perennial sod and reduced tillage results in increased soil organic matter and reduced soil compaction. Also, the use of multi-crop systems including plants with different rooting depths can help break compacted soil layers.

To reduce the likelihood of high bulk density and compaction:

- Soil disturbance and crop production activities must be minimized when soils are wet.
- Designated field roads or rows must be used for equipment traffic.
- The number of trips across the area should be reduced.
- Practices that maintain or increase soil organic matter should be used.

Conservation practices resulting in bulk density favorable to soil function include:

- Conservation crop rotation
- Cover crop
- Deep tillage
- Prescribed grazing
- Residue and tillage management

13.2. Particle Density

The weight of an individual soil particle per unit volume is called *particle density*. The average value for particle density is 2.66 g cm^{-3}. In other words, a soil particle that is 1 cubic centimeter in volume weighs 2.66 g. In comparison, water has a density of 1 g cm^{-3} and organic matter has a density of 0.8 g cm^{-3}. Particle density is important in our understanding and determination of other physical properties including bulk density and porosity.

13.3. Porosity

Porosity or pore space refers to the volume of soil voids that can be filled by water and/or air. It is inversely related to bulk density. Porosity is calculated as a per centage of the soil volume:

$$\% \text{ Porosity} = \left(1 - \frac{\text{Bulk density (BD)}}{\text{Particle density (PD)}}\right) \times 100$$

Loose, porous soils have lower bulk densities and greater porosities than tightly packed soils. Porosity varies depending on particle size and aggregation and is greater in clayey and organic soils than in sandy soils. A large number of small particles in a volume of soil produces a large number of soil pores. In contrast, fewer large particles can occupy the same volume of soil so there are fewer pores and less porosity. As bulk density increases, compaction decreases porosity. If compaction increases bulk density from 1.3 to 1.5 Mg m^{-3}, porosity decreases from 50 % to 43 %. Aggregation also decreases porosity because more large pores are present as compared to single clay and silt particles that are associated with smaller pores. Pores of all sizes and shapes constitute the total porosity of a soil. Porosity, however, does not indicate anything about the size of pores.

13.4. Pore Size

Pore size is probably one of the most important physical features of a soil, controlling water and air movement and storage. Clays have predominantly small pores and sands have large pores. Soils are a mixture of sand, silt and clay particles, so pores come in all sizes. An ideal soil condition will have an equal number of large and small pores. Large pores allow for soil aeration which is needed for the exchange of oxygen from the atmosphere and carbon dioxide given off by plant roots and microorganisms. Approximately 10 per cent of the pores must be large enough for aeration so that root growth is not restricted. Small pores are connected to large pores that are connected to medium-sized pores. Such complex connection of pores is akin to a maze with wide, medium and narrow passageways. Also, some passageways and soil pores, may be dead ends. Within an aggregate, the pores are small. Between aggregates, pores are large. Small pores are called *micropores*, while the large ones are called *macropores*. When organic matter is added, the number of macropores increases. These increases are due to the increase in aggregation, decay of root channels and creation of earthworm channels. Macropores are crushed when a soil is compacted. Tillage tends to increase macropores in the short-term, but reduces the number of macropores in the long-term because of the loss of aggregation and severing of earthworm channels.

13.5. Permeability of Compacted Soils

Soil permeability represents the ease with which air, water or plant roots penetrate or pass through soil. The degree of permeability of compacted soil is related to the rate at which water moves through soil after it has been compacted. If a coarse-grained soil, after compaction, has large continuous pores, it transmits water rapidly and is said to have a high permeability. Fine-grained soils contain very small continuous pores and a compacted fine-grained soil will transmit water very slowly and has low permeability.

Soils with good compaction characteristics can be compacted to a high degree with minimum effort. Soil material with a plasticity index close to 16 per cent has the best compaction characteristics. For each soil, there is an optimum moisture content that will permit that soil to be compacted to the maximum degree with the least effort. Optimum moisture content for various types of soils are as follows:

Soil	Optimum moisture content for compaction (range in %)
Clayey sands, sand-clay mix	11 - 10
Sand-silt-clay mix with plastic, silt + clay fraction	15 - 11
Inorganic silt, clayey silt	24 - 12
Inorganic clay	24 - 12
Organic silt	33 - 21
Inorganic clay, highly plastic	36 - 19
Organic clay	45 - 21

Note: The optimum moisture content is usually 2-3 per cent less than the plastic limit of the soil.

13.6. Compressibility

Compressibility is the degree to which a soil mass decreases in volume when supporting a load. The lowest compressibility is in coarse-grained soils where particles are in contact with each other. It increases as the proportion of small particles increases and reaches the highest in fine-grained soils which contain organic matter. The following are a few examples of compressibility for various soils:

- Gravels and sands are practically incompressible. If a moist mass of these materials is subjected to compression, there is no significant change in their volume.
- Clays are compressible. If a moist mass of clay is subjected to compression, moisture and air may be expelled, resulting in volume reduction which is not immediately recovered when the load is removed.

Fine-grained soils that contain at least 50 per cent of silt + clay may be listed in three classes of compressibility on the basis of their liquid limit. They are as follows:

- Low compressibility: LL smaller than 30
- Medium compressibility: LL from 30 to 50
- High compressibility: LL greater than 50

In general, compressibility is approximately proportional to the plasticity index (*refer* Section 12.6). The greater the PI, the greater is the compressibility of the soil.

13.7. Shrink-Swell Potential

The shrink-swell behaviour of a soil is that quality which determines its volume change under varying moisture conditions. Some soils will shrink when dry and swell when wet. The volume change of the soil mass is a function of the amount of moisture change as well as the amount and kind of clay present in the soil. Loamy sand, sand and kaolinite clay have low shrink-swell potential, while montmorillonite clay has high shrink-swell potential.

13.8. Shear Strength

The shear strength of a soil is the relative resistance of that soil to sliding when supporting a load. The highest resistance to sliding occurs in soils that are composed of clean gravel with less than 5 per cent silt + clay. Shear strength of soils decreases as the proportion of fine particles increases. It is lowest in fine-grained organic soils. For instance, it is important to remove all the organic soil to decrease the possibility of sliding when building a dam.

13.9. Measurement of Soil Density

Accurate measurements of soil bulk density are required by workers in all branches of Soil Science. Numerous methods are available for soil bulk density measurements and these can be grouped into two: those which deal with disturbed soil in the laboratory and those which deal with the soil in its natural state in the field. Since the weighing of soil does not present any difficulty, it is in the determination of soil volume that differences among methods arise. An example of the first kind of method is dipping a large aggregate of soil of known weight in paraffin wax and determining its volume by water displacement (*Shaw, C. F. 1917. A method for determining the volume of soils in field condition. J. Amer. Soc. Agron. 9: 38-42*).

An example for the second kind of method is the core cutter method which is used when coarse fragments (particles with diameter > 2 mm) occupy less than 25% by volume. At sites where lots of coarse particles and/or tree roots are found, the core method is difficult to use. The core cutter contains a steel cylinder which is driven into the soil with blows from a drop hammer. Then the cylinder containing an undisturbed soil core is removed and trimmed to the end with a knife. The volume of the core is then calculated from its length and diameter. The weight of this soil core is then determined after drying in an oven at 105 °C for about 18-24 hours.

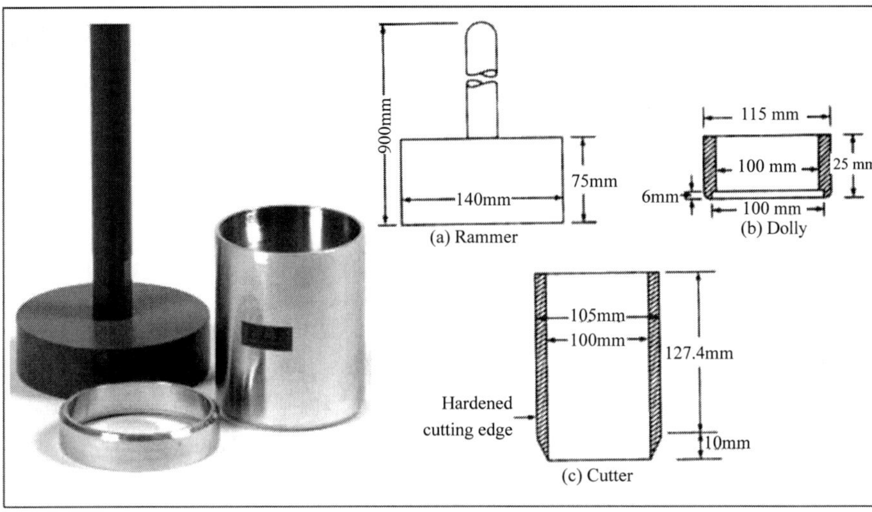

Fig. 13.2: Soil core cutter assembly (rammer, dolly and cutter)

While the methods described above necessitate removal of soil from the field for density measurements, methods are also available which determine soil bulk density *in situ* under field conditions, without having to remove soil from the field. **Density probes** use a gamma source, mainly of ^{226}Ra or ^{137}Cs, with either GM tubes or a scintillation crystal for the detection of the gamma photons, which are either *back-scattered* from, or *transmitted* through, the soil (Fig. 13.3). The most commonly used instruments (depth and surface probes) are back-scattering devices, as these are simpler to handle and cheaper in cost.

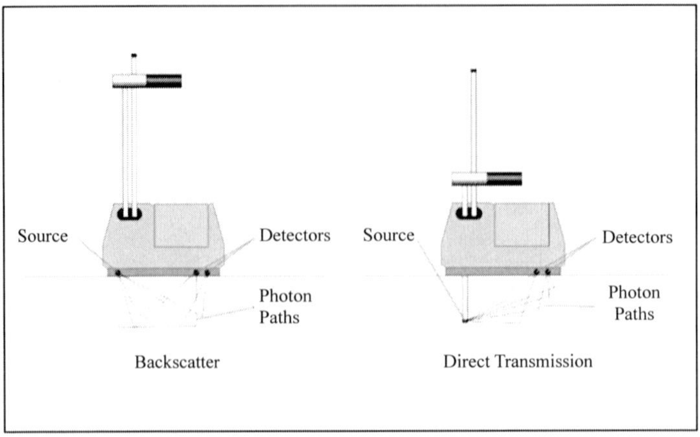

Fig. 13.3: Gamma density probes

The measurement of density is based on the known interaction of gamma rays and the orbital electrons of atoms. Gamma photons, having an energy less than 1 MeV and emitted by a radioactive source that is placed in the soil, will interact with the surrounding electrons by the Compton effect and/or photoelectric absorption. As the number of electrons per unit volume of soil (density) is increased, the Compton scattering power of the medium increases proportionately. With each scattering process, however, the gamma photon loses some of its energy. Thus, although an increase in electron density of the scattering medium (soil) increases the probability of multiple scattering of the gamma photon, the probability that the gamma photon will suffer photoelectric absorption, before it can reach the detector, is also increased. The combined effect of these two probabilities is that a smaller number of gamma photons will reach the probe as the surrounding soil becomes denser. Since the number of electrons present per unit volume of soil is approximately proportional to the density of the soil, the number of gamma rays scattered back and detected per unit of time is an inversely correlated measure of density. Since absorption of gamma photons by hydrogen is about twice as high per nucleon as other common elements in soil, this would lead to an error in bulk density caused by variation in water content. However, this error is usually reduced in magnitude, either by a circuitry technique that discriminates against low-amplitude pulses coming from the detector, or by a judicious use of lead screening to attenuate low-energy radiation reaching the detector. The density probe measures the wet density of the soil; therefore, the weight of water should be subtracted if dry density data are desired.

Owing to the importance of measuring the distribution of soil moisture in the top soil layer, a **two-well gamma probe** technique is used for this purpose. Two small access tubes, about 50 cm long, are placed in the soil, so that the distance separating them is fixed and well defined. A gamma-radiation source is placed in one tube and a detector is lowered to the same depth in the other tube. Using a small detector and a collimated (point) radiation source, the pathway of the gamma rays can be restricted to a rather thin soil layer of about 5 cm height. A portion of the emitted gamma photons is absorbed by the soil sample and the water molecules it contains, and another portion reaches the detector. The number of gamma photons that reaches the detector decreases as the moisture content of the soil sample increases. If the detector is connected to a counting unit, the indicated count-rate could be transformed to a measure for the moisture content of a certain well-defined soil layer with the aid of a calibration chart.

14

Soil Colour

Soil colour is the most obvious characteristic of soil. It has no direct effect on plant growth; however, it has indirect effect through temperature, moisture. Soil colour can indicate the climatic conditions under which it has developed and can be related to its parent material. It can also reflect upon its age, and the temperature and moisture characteristics of the climate. Thus, cooler regions tend to have grayish to black topsoils, due to the accumulation of humus. In moist warm regions, soils tend to be more yellowish-brown to red. Arid soils tend to be light in color (due to low organic content) and primarily express the color of their mineral content. Practically all colours occur in soils - white, red, brown, gray, yellow, black, etc. Predominantly soil colours are not pure, but mixtures. Frequently two or three colours occur, called as "mottling".

14.1. Causes for Soil Colour

Soil colour is a composite of the colour of its components. Organic matter (humus), manganese and iron are the primary agents for soil colour. Humus is black or brown; quartz is white. The dark color of many productive soils is due to organic matter. Iron oxides are red, brown or yellow; reduced iron is blue-green. Red soils contain oxidized iron. Yellow soils contain hydrated iron. In gray soils chemical reduction of iron and/or manganese due to wetness and lack of oxygen are evident.

Wet or moist soils are darker than dry soils. Yellow and gray coloration found in the subsoils that remain wet for some portion of the year serve as an important indicator of natural drainage conditions. In the description of soil series, these colors are designated by the term "mottles". In sloping land that has been eroded excessively, subsoil horizons may be lighter in color and possess little organic matter.

14.2. Significance of Soil Colour

Colours can say much about a soil. Generally, dark soils higher in productivity: Black > brown > red > yellow > white. Colours serve to study soil genesis. In "young" soils colour indicates parent material. In "mature" soils colour indicates climate. Mottles indicate intermittent oxidation, reduction (temporary excess of water, lack of aeration). Practically, all soil profiles reveal change of colour from one horizon to the next. In soil taxonomy, soil colour is used: Chernozem (black); Podzol (ash-gray); "alb" (white); "ochr" (light-coloured); "umbre" (dark).

14.3. Quantification of Soil Colour

Soils reflect light of a great variety of wavelengths. Human eye observes such polychromatic light as a definite colour impression. We may distinguish many colours, but may not describe them accurately. The reasons are:

- Every person has different impression of a given colour.
- Colour depends on object's own properties & quality of light it reflects.
- Our colour vocabulary is very confused.

Thus, quantification of soil colour is necessary.

14.4. Colour Disc

A colour disc is made up of sections of basic soil colours – red, yellow, black, white. The width of sections can be varied. The disc is spun & resultant colour compared to that of the soil (Shaw, 1935). The segments are adjusted till the composite colour matches the colour of soil. Colour described in terms of percentage of area of standard colours used in the disc.

14.5. Munsell Colour System

Albert H. Munsell (1905), a scientific thinker and an artist, developed a system of colours that could be used across many disciplines. Munsell color system is set on a three-dimensional model. Each color has three qualities or attributes:

1. Hue – dominant spectral colour such as red, orange, yellow, etc.
2. Value – the relative lightness or darkness of a colour
3. Chroma – the purity or brilliance of a colour

A. H. Munsell

The Munsell Color system is set up as a numerical scale with visually uniform steps for each of the three-color attributes. In Munsell color notation, each color has a logical and visual relationship to all other colors.

Hue

Hue is that attribute of a color by which we distinguish one colour from the other. There is a natural order of hues: red, yellow, green, blue, purple.

Paints of adjacent colors in this series can be mixed and a continuous variation can be obtained from one to the other. For instance, red and yellow may be mixed in any proportion to obtain all the hues from red through orange to yellow. The same is true for yellow and green, green and blue, blue and purple, and purple and red. This series returns to the starting point, so it can be arranged around a circle (Fig. 14.1). Munsell called red, yellow, green, blue, and purple as "principal hues" and placed them at equal intervals in a circle. He inserted five intermediate hues: yellow-red, green-yellow, blue-green, purple-blue and red-purple, making totally ten hues. For brevity, he used the initials as symbols to designate these ten hues: R, YR, Y, GY, G, BG, B, PB, P, and RP.

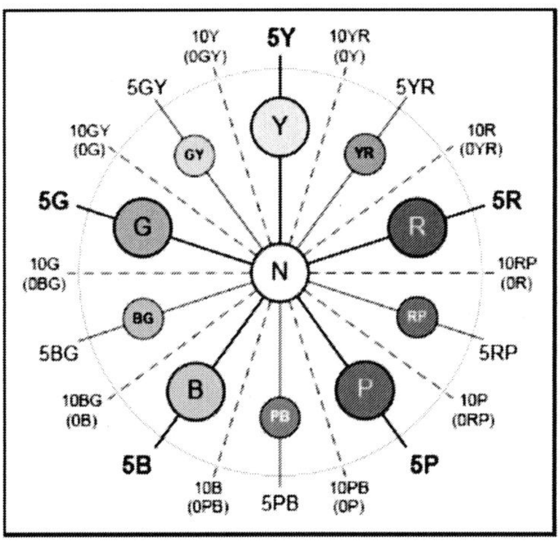

Fig. 14.1: The Munsell hue circle

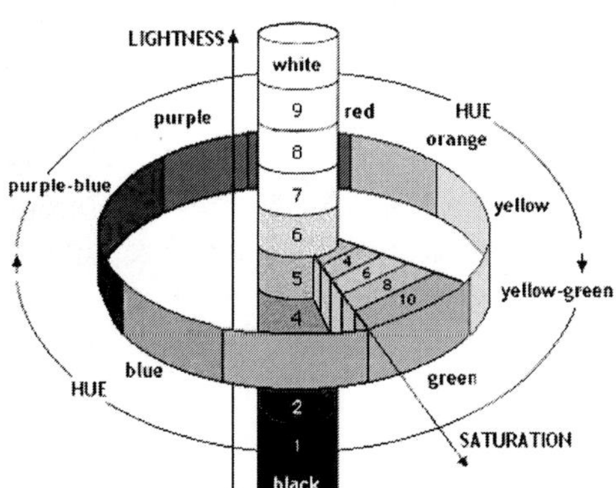

Fig. 14.2: The hue, value and chroma scales of Munsell colour system

Value

Value indicates the lightness of a color. The scale of value ranges from 0 (pure black) to 10 (pure white). Black and white, with juxtaposed grays, are called, "neutral colors" (Fig. 14.2). They have no hue. Colors having a hue are called "chromatic colors." The scale of value applies to both chromatic and neutral colors.

Chroma

Chroma is the degree of departure of a color from the neutral color of the same value. Colors of low chroma are often called, "weak," while those of high chroma are "highly saturated," "strong," or "vivid".

When a vivid red paint is mixed with a gray paint of the same value, starting with gray and gradually adding red, a little at a time, until the vivid red color is obtained, the series of gradually changing colors would exhibit increasing chroma.

Determination of Soil Colour

This done by a direct comparison of soil with colour chips (Fig. 14.3) in the Munsell soil colour chart. In the chart, all colors on a given card are of a constant hue, designated by a symbol in the top right-hand corner. Vertically, the colors are progressively lighter by visually equal steps and their value increases. Horizontally, they increase in chroma to the right and become grayer to the left. The value and chroma of each color in the card are mentioned immediately beneath the color, the first number being the value and the second being the chroma. Thus, the colors arranged in the chart form three scales: (1) radial, or from one card to the next, in hue; (2) vertical in value; and (3) horizontal in chroma.

Thus, each colour chip is designated by a numerical notation that includes hue, value and chroma as described above. (The notation is written thus: Hue, followed by a space and then value/chroma. For example: 10YR 5/2. In this notation, 10YR represents hue, 5 represents value and 2 represents chroma.) There are circular holes between chips to put peds and compare colour. A description of soil colour in words is also given in the Munsell colour chart, corresponding to each notation. The colour description for above example 10YR 5/2 is grayish brown.

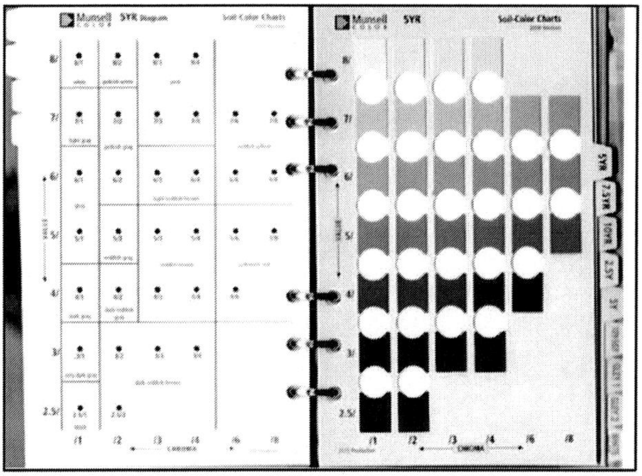

Fig. 14.3: Munsell soil colour chart

To determine soil color, a fresh ped (a unit of soil structure) interior is used to eliminate the error of describing the color of the surface coating of the ped. Since a moist ped will appear darker than a dry ped, description of soil colour must qualify whether the soil was dry or moist when the colour determination was done. For example: 10YR 5/1 (dark gray) when dry.

15

Soil Temperature and Soil Air

15.1. Soil Temperature

Soil temperature affects plant growth directly; influences soil moisture, aeration, structure, microbial activity, decomposition of plant residues and availability of plant nutrients. Intermediate temperatures, near optimum for each crop, are the desirable temperatures for crop growth. Extreme temperatures are undesirable, lest crop growth would be retarded and weeds and germs may be favoured. Moderate daily variations are desirable as they lead to soil moisture distribution and aeration.

15.1.1. Factors Affecting Soil Temperature

The mean annual soil temperature is about 1°C higher than the mean annual air temperature. Soil temperature is influenced mainly by climatic conditions. The factors which affect the transfer of heat through atmosphere from the Sun also affect the soil temperature. The following are the factors that affect soil temperature:

Environmental Factors

- Solar radiation
- Conduction of heat from atmosphere
- Condensation
- Evaporation
- Rainfall
- Vegetation

The magnitude of heat received from the Sun at Earth's surface is 2 cal cm^{-2} min^{-1}. But the magnitude of heat transmitted into soil is much smaller and it depends on the angle of incident radiation, latitude, season, time of the day, steepness and direction of slope and altitude. Insulation by air, water vapour, clouds, dust, smog, snow, vegetation, mulch, etc., moderates the amount of heat transferred into soil.

Soil Factors

- Thermal (Heat) capacity of soil

  The amount of energy required to raise the temperature by 1°C is called *heat capacity*. When it is expressed per unit mass (calories per gram), then it is called as *specific heat*. The specific heat of water is 1.00 cal g^{-1}, whereas the specific heat of a dry soil is around 0.2 cal g^{-1}. Increasing water content in soil increases the specific heat of the soil and, therefore, a dry soil heats up quickly than a moist soil.

- Thermal conductivity

  Thermal conductivity denotes the movement of heat in soils. In soil, heat is transmitted through conduction. Heat passes from soil to water about 150 times faster than from soil to air. Thus, the movement of heat will be more in wet soil than in dry soil where the pores will be filled with air. A dry and loosely packed soil will conduct heat slower than a wet and compact soil.

- Biological activity

  When soil animals, microbes and plant roots respire, heat is evolved. The more the biological activity, the more will be the soil temperature.

- Structure, texture, moisture

  Compact soils have greater thermal conductivity than loose soils. Mineral soils have higher conductivity than organic soils. Moist soils will have uniform temperature over depth because of good conductivity than dry soils.

- Soluble salts

  Soluble salts indirectly affect soil temperature by influencing the biological activities, evaporation, etc.

15.1.2. Soil Temperature Management

Scope for controlling soil temperature is limited. However, relatively small changes induced can have pronounced effects. Changes in soil moisture content and structure can alter soil temperature. Most effective way is by use of mulch (crop residues, plastics, etc.). Vegetation keeps the soil cool.

15.1.3. Soil Temperature Measurement

The instruments used for soil temperature measurement are mercury thermometers, thermocouples and thermistors. For precise temperature measurements, thermocouples are preferred because of their quick response to sudden changes in temperature and ease of automation. The international standard depths for temperature measurements are 10, 20, 50 and 100 cm.

Fig. 15.2: Soil thermometer

15.2. Soil Air

The composition of soil air is different from that of the atmosphere. The metabolic activity of plant roots, microbes, soil fauna and the conditions of soil all affect the composition of soil air. For example, the concentration of carbon dioxide in soil (between 0.3 and 3%) is often several hundred times higher than the 0.03% in the atmosphere (Table 15.1). In extreme cases, oxygen can be as low as 5-10%, compared to 20% in the atmosphere. Soil air has a higher moisture content than the atmosphere, with relative humidity approaching 100% under optimum conditions. Humidity is not as variable in soil as it is in the atmosphere. The amount and composition of air in soil are dynamic and to a large degree are determined by soil moisture and activity of soil organisms.

Table 15.1: Comparison between soil air and atmospheric composition

Component	Soil air (%)	Atmosphere (%)
Nitrogen	79.2	79.0
Oxygen	20.6	20.9
Carbon di oxide	0.25	0.03

15.2.1. Movement of Gases within Soil

Exchange of gases between soil and atmosphere is brought about by two mechanisms:

1. **Mass flow (Convection) of air:** The moving force here is a *gradient of total gas pressure* and it results in the entire mass of air streaming from a zone of higher pressure to one of lower pressure. Mass flow of air is much less important than diffusion, except perhaps in layers at or very near the soil surface.

2. **Diffusion:** The moving force here is gradient of partial pressure of *any constituent member* of air to migrate from a zone of higher pressure to a zone of lower pressure, even while air as a whole may remain stationary.

In other words, by diffusion each gas moves in a direction as determined by its own partial pressure.

The availability of many nutrients is influenced by soil aeration. Many microorganisms that release plant nutrients to the soil particularly need soil air. Most plants require maintenance of adequate soil oxygen for their growth and root proliferation.

Air will fill soil pores as water drains or as water is removed from soil pore by evaporation or root absorption. The network of pores within the soil aerates the soil. When water enters soil pores, this aeration network is blocked. Soil air and soil water are very dynamic parts of a soil, but both are mutually exclusive: an increase in soil water content causes a reduction in soil aeration. Likewise, reducing soil water content leads to an increase in soil aeration.

Since plant roots require water and oxygen, maintaining a just balance between root aeration and soil water availability is an intricate aspect of managing crop plants. Soil air is very different from the above-ground atmosphere. Since the soil contains high amounts of carbon dioxide, oxygen levels may often become limited. Since plants need oxygen to live, it is imperative to allow proper aeration in the soil.

15.2.2. Soil Porosity

Soil porosity (f) is the ratio of pore volume (V_f) to total soil volume (V_t)

$$f = V_f / V_t$$

It is generally between 30-60%. Porosity does not tell about the relative amounts of large and small pores and should be interpreted with caution. Generally, high porosity (e.g. 60%) is an indicator of lack of compaction and good soil conditions.

16

Soil Water

Water is the elixir of life. It is required for the evapo-transpiration of growing plants. Water acts as a solvent and, together with dissolved nutrients, makes up soil solution from which plant roots derive nutrients. Soil water content controls soil air and soil temperature.

16.1. Forces of Water Retention

The forces responsible for water retention in soils are adhesion and cohesion. Adhesion refers to the attraction of solid surfaces for water. Cohesion refers to the attraction of water molecules for each other.

The force with which the soil particles retain water is defined in terms of pF scale (Schofield, 1935). pF is defined as "the logarithm to the base 10 of the numerical value of the negative pressure of soil moisture expressed in centimetres of water."

pF = log h, where h is the height in centimetres of a column of water which would give a pressure.

16.2. Retention of Soil Moisture

Based on the magnitude of moisture held by soil, physically, there are four kinds of soil moisture:

- Maximum retentive capacity
- Field capacity
- Wilting coefficient
- Hygroscopic coefficient

16.2.1. Maximum Retentive Capacity

As the water enters dry soil, air is displaced. Soil pores, large and small are filled with water. Continued water application leads to further downward movement and air displacement. Now, all soil pores are filled with water. Soil is now "saturated" and is at "maximum retentive capacity".

16.2.2. Field Capacity

After saturation, if water supply is cut, continued rapid downward movement of some water takes place. After a day or two, this rapid downward movement will essentially cease. The soil is then at "field capacity" (pF 2.53). Water has moved out of macropores and its place taken by air. Micropores are still filled with water; field capacity moisture is plant available.

16.2.3. Wilting Coefficient

Plants begin to show effects of reduced soil moisture. During day time they tend to wilt; renewed night time turgor. Ultimately plants remain wilted night and day. Wilting point or wilting coefficient is defined as "the soil moisture condition at which the ease of release of water to the plant roots is just too small to counter balance the transpiration losses". This corresponds to a pF of 4.18 or 15 atmospheres

16.2.4. Hygroscopic Coefficient

Hygroscopicity is the ability of soil to absorb moisture from atmosphere. It is "the amount of moisture in a dry soil when it is in equilibrium with some standard relative humidity near a saturated atmosphere (about 98%), expressed as per centage on the basis of oven-dry soil" and corresponds to pF of 4.5 or 30.5 atmospheres.

16.3. Soil Moisture Availability

The factors influencing soil moisture availability to crop plants are:

Plant factors: Rooting habits, stage and rate of growth, etc.

Climatic factors: Air, temperature, humidity, etc.

Soil factors: Soil moisture tension relations (Texture, structure, OM), salt content, soil depth, soil layering [hard pans, impervious layers, etc.]

16.4. Soil Moisture Measuring Methods and Devices

A variety of methods and devices can be used to measure soil moisture. These include the gravitational method, tensiometer, electrical resistance blocks, neutron probe, phene cells, and time domain reflectometer. Most of these methods and devices - instead of measuring soil water directly - measure a property of the soil that can be related to soil water status and are therefore are considered as indirect methods. These methods vary with respect to their ease of use, reliability, cost, and amount of labor required.

16.4.1. Gravimetric Method

In this method, soil moisture is determined by drying a weighed soil sample from the desired soil depth in an oven (for 24 hours at 220 °F), and then reweighing the dry sample to determine the quantity of water was lost. This method is simple and reliable. Unfortunately, it is not practical for scheduling irrigation because it takes a full day to dry the sample. A sandy soil which dries quickly may need irrigation even before the results of the measurement are obtained. The gravimetric method is most useful for calibrating devices used for measuring soil-water.

16.4.2. Tensiometer

A tensiometer is a sealed, airtight, water-filled tube (barrel) with a porous tip at one end and a vacuum gauge at the other, as shown in Fig. 16.1. It measures soil water suction (negative pressure), usually expressed as tension. This suction is equivalent to the force or energy that a plant must exert in order to extract water from soil. The tensiometer must be installed properly so that the porous tip is in good contact with the soil and it must be ensured that the soil water suction is in equilibrium with the water suction in the tip. The suction force in the porous tip is transmitted through the water column inside the tube and displayed as a tension reading on the vacuum gauge. Soil water tension is commonly expressed in units of bars or centibars.

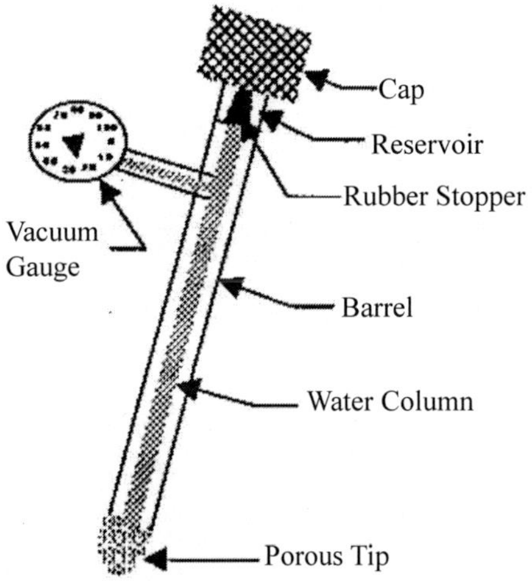

Fig. 16.1: Tensiometer

The suction at the tip is transmitted to the vacuum gauge because of the cohesive forces between adjacent water molecules. As the suction approaches approximately 0.8 bar (80 cb), the cohesive forces are exceeded by the suction and the water molecules separate. When this occurs, air can enter the tube through the porous tip and the tensiometer no longer functions correctly. This condition is referred to as *breaking tension*. Tensiometers work in the range from 0 to 0.8 bar. The suction scale on the vacuum gauge of most commercial tensiometers reads from 0 to 100 cb.

Tensiometers are best suited for use in soils that release most of their plant available water (PAW) at soil-water suctions between 0 and 80 cb. Soil textures in this category are those that consist of sand, loamy sand, sandy loam, and the coarser-textured range of loam and sandy clay loam. Many clayey and silty soils still retain over 50 per cent of their plant-available water at suctions greater than 80 cb, which is outside the working range of a tensiometer.

16.4.3. Electrical Resistance Blocks

Electrical resistance blocks consist of two electrodes enclosed in a block of porous material, as shown in Fig. 16.2. The block is often made of gypsum, although fiberglass or nylon is sometimes used. Electrical resistance blocks are often referred to as *gypsum blocks*. The electrodes are connected to insulated lead wires that extend upward to the soil surface. The block is buried in the soil at one-half the effective root depth. With the proper calibration curve, the meter reading can be related to soil moisture.

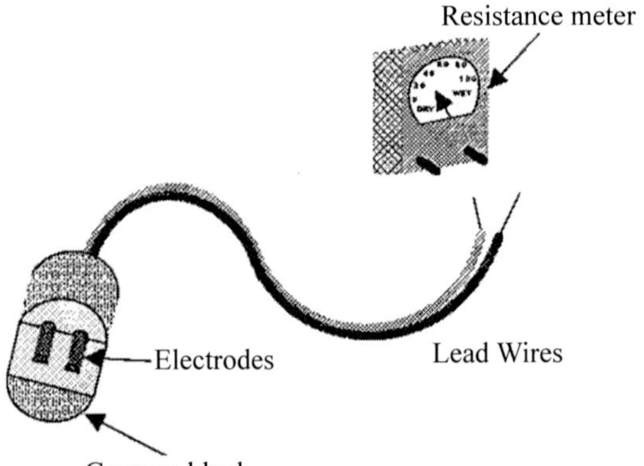

Fig. 16.2: Schematic of an electrical resistance block and meter

Resistance blocks work on the principle that water conducts electricity. If properly installed, the water suction of the porous block is in equilibrium with the soil-water suction of the surrounding soil. With the change in soil moisture, the water content of the porous block also changes. The electrical resistance between the two electrodes increases as the water content of the porous block decreases. By means of a calibration curve, the block's resistance can be related to the water content of the soil.

Because of the pore size of the material used in most electrical resistance blocks, particularly those made of gypsum, the water content and thus the electrical resistance of the block does not change dramatically at suctions less than 0.5 bar (50 cb). Therefore, resistance blocks are best suited for use in fine-textured soils such as silts and clays that retain at least 50 per cent of their plant-available water at suctions greater than 0.5 bar. Electrical resistance blocks cannot be relied upon to determine when to irrigate sandy soils where more than 50% of the plant-available water is usually depleted at suctions less than 0.5 bar.

16.4.4. Neutron Probe

The neutron probe (Fig. 16.3) uses a radiation source to measure soil-water. An empty tube (access tube) with a 2-inch inside diameter must be installed vertically in the soil at each field location where the soil-water is to be measured. The neutron method uses the principle of neutron moderation. Hydrogen nuclei have a marked propensity for scattering and slowing neutrons. During operation, high energy neutrons emitted from a radioactive source, such as ^{226}Ra-Be or ^{241}Am-Be, slow down by several elastic collisions with hydrogen nuclei of soil water (moderation). The thermal energy of neutrons is reduced to about the thermal energy of atoms in a substance at room temperature. Hydrogen, which has a nucleus of about the same size and mass as the neutron, has much greater moderating effect on fast neutrons than any other element. The thermal neutron density easily measured with a detector is calibrated against water concentration on a volume basis. If perfectly calibrated, the neutron probe is easy to use, reliable and accurate.

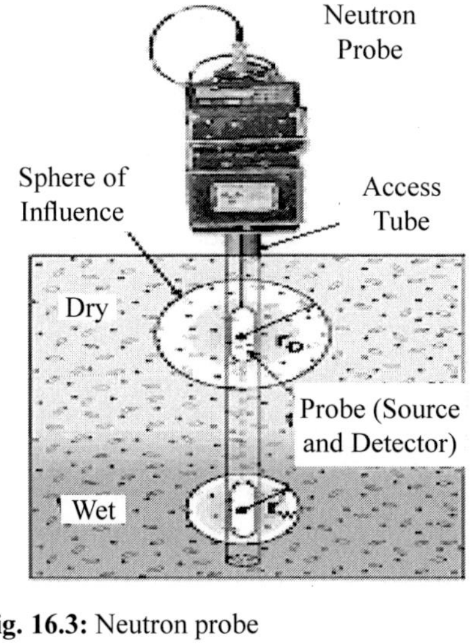

Fig. 16.3: Neutron probe

One of its advantages is that soil-water measurements can be made easily at different depths in the soil profile. Because of its cost, a neutron probe may be a viable option for operators with large acreages of irrigated land. At present, it is used by some irrigation consultants to perform the technical tasks required to schedule irrigation. Moisture readings from a neutron probe are downloaded to irrigation scheduling programmes to build a field history file and generate reports of soil moisture status and irrigation forecasts. This helps to improve crop yields and quality, determine when and how much to irrigate, monitor root development and improve irrigation efficiency.

16.4.5. Two-well Gamma Probe

Owing to the importance of measuring the distribution of soil moisture in the top soil layer, a two-well gamma probe technique is used for this purpose. Two small access tubes, about 50 cm long, are placed in the soil, so that the distance separating them is fixed and well defined. A gamma-radiation source is placed in one tube and a detector is lowered to the same depth in the other tube. Using a small detector and a collimated (point) radiation source, the pathway of the gamma rays can be restricted to a rather thin soil layer of about 5 cm height. A portion of the emitted gamma photons is absorbed by the soil sample and the water molecules it contains, and another portion reaches the detector. The number of gamma photons that reaches the detector decreases as the moisture content of the soil sample increases. If the detector is connected to a counting unit, the indicated count- rate could be transformed to a measure for the moisture content of a certain well-defined soil layer with the aid of a calibration chart.

16.4.6. Cosmic Ray Soil Neutron Sensor

Most sensors traditionally used in the measurement of soil water content monitor areas less than 1 m in radius. Therefore, a large number of sensors are required to obtain data on a large area of soil and this can be both costly and laborious. The cosmic ray soil neutron scanner (Fig. 16.4) is a device that is used to measure water content in top soil over wide areas, up to 30 hectares, and to a depth varying from 0.1 to 0.6 m, in a non-invasive, non-hazardous and continuous way.

High-energy cosmic rays from outside the solar system produce neutrons. These collide mainly with nitrogen and oxygen atoms in the Earth's upper atmosphere that break into subatomic particles such as protons and neutrons which rain through the atmosphere and continue to collide with other atoms as they fall. The neutrons that reach the earth's surface are very fast-moving. Their energy is absorbed by atoms in the environment, with hydrogen atoms absorbing the most. This absorption slows the neutrons down. As most of the hydrogen in the terrestrial environment is found in water in soil, scientists can count the number of fast neutrons in and around soils, to determine how much water is present.

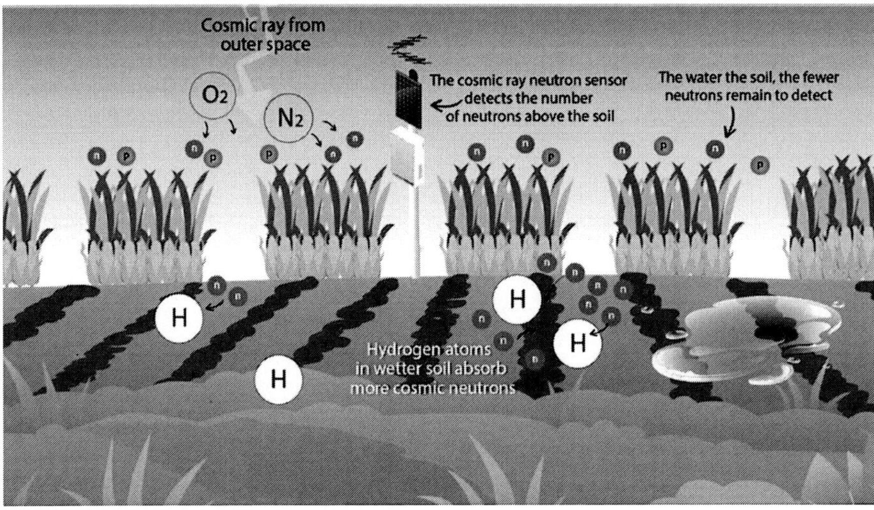

Fig. 16.4: Cosmic ray soil neutron sensor

Measurements of fast-moving neutrons in variably□saturated soils showed that neutron intensity and soil moisture content are inversely correlated, indicating the feasibility of invasive cosmic□ray neutron measurements. Because neutrons that are generated and moderated in the ground diffuse back into the air, this feasibility also applies in the case of non□invasive measurements above the ground surface.

16.4.7. Time Domain Reflectometer

The time domain reflectometer (TDR) is a device developed to measure soil-water content. Two parallel rods (or stiff wires) are inserted into the soil to the desired depth and the rods are connected to an instrument that sends an electromagnetic pulse (or wave) of energy along the rods. The rate at which the wave of energy is conducted into the soil and reflected back to the soil surface is directly related to the average water content of the soil. One instrument can be used for hundreds of pairs of rods.

17

Soil Water Potential

The phenomena related to the retention and movement of water in soils, its uptake and translocation in plants and its loss to the atmosphere are all energy related. The more strongly water is held in the soil, the greater is its energy. It means that if water is to be removed from a moist soil, work has to be done against the adsorptive forces. The water movement is always from a zone where the free energy of water is high (standing water table) to one where the free energy is low (a dry soil). This is called *soil water energy* concept.

17.1. Soil Water Potential

The free energy of soil solids for water is influenced by:

i) **Matric force:** the attraction of the soil solids for water (adsorption) which markedly reduces the free energy (movement) of the adsorbed water molecules.

ii) **Osmotic force**: the attraction of ions and solutes for water to reduce the free energy of soil solution, also called as *solute potential.* Matric and osmotic potentials reduce the free energy level of the soil water and are negative and. These negative potentials are referred to as *suction* or *tension.*

iii) **Force of gravity**: This acts on soil water, the direction being towards the earth's center, and tends to pull the water downward. This force is always positive. The difference between the energy states of soil water and pure free water is known as *soil water potential.* Total water potential (P_t) is the sum total of the contributions of gravitational potential (P_g), matric potential (P_m) and the osmotic potential (P_o).

$$P_t = P_g + P_m + P_o$$

17.2. Methods of Expressing Suction

There are two ways in which the differences in energy levels of soil water are expressed.

i) **pF Scale**: The free energy is measured in terms of the height of a column of water required to produce necessary suction or pressure difference at a particular soil moisture level. The pF, therefore, represents the logarithm of the height of water column (cm) to give the necessary suction.

ii) **Atmospheres or Bars**: This is another common mean of expressing suction. Atmosphere is the average air pressure at sea level. If the suction is very low, as in the case of a wet soil containing the maximum amount of water that it can hold, the pressure difference is of the order of about 0.01 atmospheres or 1 pF equivalent to a column of water 10 cm in height. Similarly, if the pressure difference is 0.1 atmosphere the pF will be 2.0. Soil moisture constants can be expressed in term of pF values. A soil that is saturated with water has pF 0 while an oven dry soil has a pF 7.

17.3. Soil Moisture Constants

Earlier classification differentiated soil water into three classes, namely, gravitational, capillary and hygroscopic water. The hygroscopic and capillary waters are in equilibrium with the soil under given condition. The two equilibrium points when the soil contains the maximum amount of hygroscopic and capillary waters, are respectively the hygroscopic coefficient and the maximum capillary capacity. The amount of water that a soil contains at each of these equilibrium points is known as *soil moisture constant.*

The soil moisture constant represents definite soil moisture relationship and retention of soil moisture in the field. However, the three classes of water (gravitational, capillary and hygroscopic) are very broad and do not explain accurately the true soil - water relationships that exist under field conditions. Hence, two more soil constants, *viz., field capacity* and *wilting coefficient* were introduced to describe the soil-plant-water relationship as it exists under field conditions.

17.3.1. Field Capacity

When water is applied to the surface of a soil, there is downward movement of water and all macro- and micro-pores are filled up. The soil is then said to be saturated and is at maximum water holding capacity or maximum retentive capacity which is the amount of water held in soil when all the pores are filled.

A certain time after the application of water in the soil, all the gravitational water is drained away (from macro-pores) and then the wet soil is almost uniformly moist. The amount of water held by the soil at this stage is known as *field capacity* which is *the capacity of the soil to retain water against the downward pull of the force of gravity.* At this point only micropores or capillary pores are filled with water and plants absorb water for their use. The force with which water is held at field capacity is 1/3 atmosphere. Water at field capacity is readily available to plants and microbes.

17.3.2. Wilting Coefficient

As the moisture content continues to fall, a point is reached when the water is so firmly held by the soil particles that plant roots are unable to tap it anymore. The plant then begins to wilt. At this stage even if the plant is kept in a saturated atmosphere it does not regain its turgidity and wilts unless water is applied to the soil. The stage at which this occurs is termed the *wilting point* and *the amount of water held by the soil at wilting point is known as wilting coefficient*. It represents the point at which the soil is unable to supply water to the plants. At wilting coefficient, water is held with a force of 15 atmospheres.

17.3.3. Hygroscopic Coefficient

The hygroscopic coefficient is the maximum amount of hygroscopic water absorbed by 100 g of dry soil under standard conditions of humidity (50% relative humidity) and temperature (15 °C). This tension is equal to a force of 31 atmospheres. Water at this tension cannot be available to plants, but may be available to certain bacteria.

17.3.4. Available Water Capacity

The amount of water required to be applied to a soil at the wilting point in order to reach the field capacity is called *available water*. The water supplying power of a soil is directly related to the amount of available water it can hold. *The difference between the amount of water at field capacity (- 0.3 bar) and the amount of water at the permanent wilting point (- 15 bars) is called available water.*

17.3.5. Maximum Water Holding Capacity

It is also known as maximum retentive capacity. It is the amount of moisture in a soil when its pore spaces both micro and macro capillary are completely filled with water. It is a rough measure of total pore space of soil. Soil moisture tension is very low, varying between 1/100th to 1/1000th of an atmosphere or pF 1 to 0.

17.3.6. Sticky Point Moisture

It represents the moisture content of soil at which it no longer sticks to a foreign object. The sticky point represents the maximum moisture content at which a soil remains friable. Sticky point moisture values vary nearly approximate to the moisture equivalent of soils.

A summary of the soil moisture and force with which it held is given in Table 17.1.

Table 17.1: Soil moisture and range of tension and pF

S. No.	Moisture class	Range of tension in atmosphere	Equivalent pF range
1.	Hygroscopic	31 to 10,000	4.50 to 7.00
2.	Hygroscopic coefficient	31	4.50
3.	Wilting point	15	4.20
4.	Capillary	1/3 to 31	2.54 to 4.50
5.	Moisture equivalent	1/3 to 1	2.70 to 3.00
6.	Field capacity	1/3	2.54

18

Soil Water Movement

Water can move in the liquid and gaseous phases:

1. Liquid water

   i) *Saturated flow*: Most pores are filled with water; water in this condition is tension free

   ii) *Unsaturated flow*: Pores are partially filled with air; the water is under tension.

2. Water vapour

   i) *Diffusion*: Water vapour moves by diffusion as a result of vapour pressure differences (partial pressure)

   ii) *Mass flow*: Water vapour flows in a mass with other gases of the system in response to differences in total pressure.

18.1. Saturated Flow

Saturated flow occurs when the soil pores are completely filled with water. This water moves at water potentials larger than –1/3 bar. Saturated flow is water flow caused by gravitational pull, beginning with *infiltration* which is water movement into soil when rain or irrigation water is on the soil surface. When the soil profile is wetted, the movement of more water through the wetted soil is termed as *percolation*.

Hydraulic conductivity can be expressed mathematically as:

$$V = kf$$

Where, V = Total volume of water moved per unit time

f = Water moving force (driving force)

k = Hydraulic conductivity of soil

Darcy's law defines the vertical flow rate of water through soil. The law states that "*the rate of flow of a liquid through a porous medium is proportional to the hydraulic gradient in the direction of flow of the liquid*".

$$Q_W = - k (d_w) At / d_s$$

Where,

Q_W = Quantity of water in cm^3

k = rate constant or hydraulic conductivity (cm s^{-1})

d_w = Water height (head) in cm

A = Soil area (cm^2)

t = Time (s)

d_s = Soil depth (cm)

18.2. Unsaturated Flow

Unsaturated flow is the flow of water held with potentials lower than -1/3 bar, when soil pores are partially filled with water. Water will move towards the region of lower potential. In a uniform soil this implies that water moves from wetter to drier areas. The water movement may be in any direction, but the rate of flow is greater as the water potential gradient (the difference in potential between wet and dry) increases and as the size of water filled pores also increases. Two forces that are responsible for this movement are adhesion (the attraction of soil solids for water) and capillarity. Under field conditions this movement occurs when the soil macropores (noncapillary pores) are filled with air and the micropores (capillary pores) are filled partly with water and partly with air.

18.3. Water Vapour Movement

The movement of water vapour from soils takes place as (a) internal movement and (b) external movement. In internal movement, the change from the liquid to the vapour state takes place within the soil, that is, in the soil pores. The external movement phenomenon occurs at the land surface and the resulting vapour is lost to the atmosphere by diffusion and convection. The movement of water vapour by diffusion mechanism takes place from one soil area to the other, depending on the vapour pressure gradient which is the moving force. This gradient is simply the difference in vapour pressure between two points a unit distance apart. The greater this difference, the more rapid is the diffusion and the greater is the transfer of water vapour during a unit period.

Two soil conditions that mainly affect the water vapour movement, *viz.*, moisture regimes and thermal regimes. In addition to these regimes, various other factors like organic matter, vegetative cover, soil colour, etc., also affect the movement of water vapour. The movement takes place from a moist soil having high vapour pressure to a dry soil with low vapour pressure. Also, the vapour movement takes place from warmer soil regions to cooler soil regions. In dry soils some water movement takes place in the vapour form and this entails practical implications in supplying water to drought resistant plants.

18.4. Entry of Water into Soil

18.4.1. Infiltration

Infiltration is the downward entry of water into the soil surface. It is a surface characteristic and hence primarily influenced by the condition of surface soil. Where the soil surface has vegetative cover the infiltration rate will be more than in bare soil. Warm and dry soils absorb more water than cold or wet ones. Coarse surface texture, granular structure and high organic matter content in surface soil, all aid in increased infiltration. The factors affecting infiltration are clay minerals, soil texture, soil structure, moisture content, vegetative cover and topography.

18.4.2. Percolation

The movement of water through a column of soil is called *percolation*. It gains importance in two contexts: (i) this is the only source of recharge of ground water which can harnessed through wells for irrigation; (ii) percolating waters leach plant nutrients down, often out of reach of plant roots. In dry regions, percolation is practically negligible and in high rainfall regions it is high. Sandy soils have greater percolation as compared to clayey soil. Vegetation and high water table reduce the percolation losses.

18.4.3. Permeability

Permeability indicates the relative ease of movement of water within the soil. The characteristics that determine how fast air and water can move through the soil defines the property of permeability. The term *hydraulic conductivity* which refers to the readiness with which a soil transmits fluids through it is also used.

18.5. Hysterisis

The word *hysteresis* is derived from the Greek, meaning "a deficiency'. In Physics, systems that take different pathways for reversing the same process have the characteristics of hysteresis. In soil also, sorption (wetting of a dry soil) and desorption (drying of a wet soil) are two contrasting phenomena that follow different paths such that the moisture held in a particular metric suction in sorption process and desorption process are *not* equal. This effect is called as *hysteresis*.

The main reason for the hysteresis phenomenon in soils is that the soil pores are geometrically heterogenous as they have different size configurations. Usually, larger pores are connected with very narrow connecting capillary pores. This difference in path diameter leads to deviations in the wetting and drying paths of the soil system. The moisture content is always low during sorption and high during desorption. Shrinking and swelling phenomena affect pore size on a micro-basis as well as on the basis of overall bulk density. Thus, hysteresis phenomenon occurs due to factors like shape and size of soil pores and their interconnection with each other, nature of soil colloids, bulk density of soil and entrapped air. The most important factor influencing hysteresis is the entrapment of air in the soil under rewetting condition which clogs some pores and prevents effective contact between others. Another reason may be the difference in the contact angle in wetting and drying phases which decreases as a surface is wetted, from $\theta = 180$ in fully dry condition to $\theta = 0$ in fully wet condition. A given amount of water will exhibit greater suction in the desorption curve due to the increasing contact angle.

18.6. Soil Moisture Characteristic

The magnitude of water remaining in the soil is a function of soil water tension. The curve showing the relationship between soil water potential ψ and soil water content θ is called the "*soil moisture characteristic*" or the "*retention curve*" (Fig. 18.1).

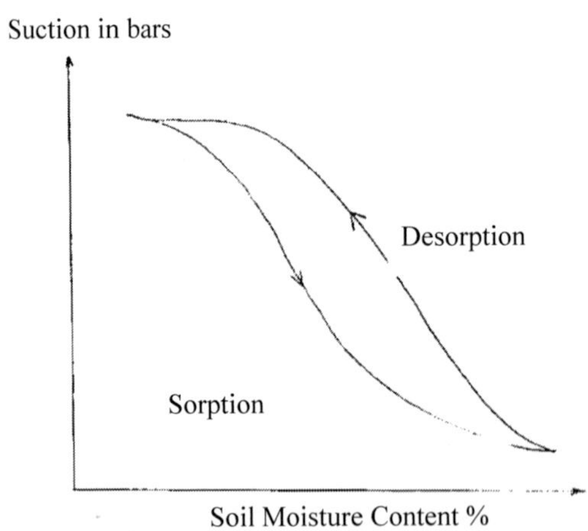

Fig. 18.1: Soil moisture characteristic

This type of characteristic curve may be generated by simply plotting the tensiometer reading of ψ and volumetric water content θ on that particular suction. It has two methodologies, *viz.*, progressively wetting a dry soil (sorption curve) and drying a wet soil (desorption curve). The sorption curve and desorption curves do not follow the same reversible path. However, the sorption and desorption curves coincide with each other at higher and lower ranges of suction force.

19

Soil Colloids

Soil Colloids

A colloidal system is a two-phase heterogenous system in which one material in a very finely divided state is dispersed through the second phase. Solid in liquid (dispersion of clay in water) and liquid in gas (fog or clouds in atmosphere) are examples. The clay fraction of the soil contains particles less than 0.002 mm in size. Particles less than 0.001 mm size possess colloidal properties and are known as *soil colloids*.

19.1. General Properties of Soil Colloids

1. **Size:** The inorganic and organic soil colloids are extremely small in size, smaller than 2 μm in diameter. These particles cannot be seen using an ordinary light microscope, but only with an electron microscope.

2. **Surface Area:** By virtue of their size, all soil colloids have a larger external surface area per unit mass. The external surface area of one gram of colloidal clay is 1000 times the surface area of one gram of coarse sand. Certain silicate clays possess extensive internal surfaces occurring between plate like crystal units that make up each particle, often greatly exceeding the external surface area. The total surface area of soil colloids ranges from 10 $m^2 g^{-1}$ for clays with only external surfaces to more than 800 $m^2 g^{-1}$ for clays with extensive internal surfaces as well. The colloidal surface area in the upper 15 cm in a hectare of a clay soil could be as high as 700,000 $km^2 g^{-1}$!

3. **Surface Charges:** Soil colloids carry negative and/or positive charges on both external and internal surfaces. Most organic and inorganic soil colloids carry a net negative charge. When an electric current is passed through a suspension of soil colloidal particles they migrate to anode, indicating that they carry a negative charge. The magnitude of the charge is known as *zeta potential*. The presence and intensity of the charges influence the attraction or repulsion of the particles towards each other, thereby influencing both physical and chemical properties. The sources of negative charge on clays come from: (1) ionizable hydrogen ions and (2) isomorphous substitution.

Ionizable hydrogen ions: These are hydrogen from hydroxyl (OH) ions on clay surfaces. The -Al-OH or -Si-OH portion of the clay ionizes the H and leaves an un-neutralized negative charge on the oxygen ($-AlO^-$ or - SiO^-). The extent of ionized hydrogen depends on solution pH and hence these negative charges are *pH dependent charges*. The ionization is greater in alkaline solutions.

Isomorphous substitution: Isomorphous substitution is the substitution in the clay crystal structure of a cation of higher valence by another cation of lower valence but similar in size. In clay crystals some ions fit exactly into mineral lattice sites because of their convenient size and charge. Clays have predominantly Si^{4+} in tetrahedral sites and Al^{3+} in octahedral sites. Other ions present in large amounts during clay crystallization can replace some of the Al^{3+} and Si^{4+} cations. Common substitutions are the replacement of Si^{4+} by Al^{3+} and Al^{3+} by Fe^{2+}, Mg^{2+} or Zn^{2+}. As the total negative charge from the anions (oxygen) remains unchanged, the lower positive charge of the substituting cations results in excess negative charges on clay crystals.

4. **Adsorption of Cations**: As soil colloids possess negative charge, they attract and attach cations like H^+, Al^{3+}, Ca^{2+} and Mg^{2+} on colloidal surfaces. This gives rise to an ionic double layer.

   Isomorphous substitutions in the colloidal particles render the external and internal layers of clay minerals negatively charged and these surfaces act as huge anions, forming the inner layer of the double layer. The outer layer is made up of a swarm of loosely held cations adsorbed to the negatively charged surfaces.

5. **Adsorption of Water**: A large number of water molecules are associated with soil colloidal particles. Some water molecules are attracted to the adsorbed cations rendering the cations hydrated. Other water molecules are bound in the internal surfaces of the colloidal clay particles. These water molecules play a critical role in determining both the physical and chemical properties of soil.

6. **Cohesion**: Cohesion is the attractive force between similar molecules or materials. Cohesion indicates the tendency of clay particles to stick together. This tendency is because of the attraction of clay particles towards water molecules held between them. Upon wetting of colloidal substances, water first adheres to individual clay particles and then brings about cohesion between two or more adjacent colloidal particles.

7. **Adhesion**: Adhesion is the attractive force between different molecules or materials. Adhesion refers to the attraction of colloidal materials to

the surface of any other body or substance with which it comes into contact.

8. **Swelling and Shrinkage**: Some soil clay colloids belonging to smectite group (like montmorillonite) swell when wet and shrink when dry. Soils high in smectite clay (Vertisols) often show wide and deep cracks after a prolonged dry spell. These cracks first allow rain to percolate rapidly. Later, because of swelling, the cracks will close and become impervious. On the other hand, soils dominated by kaolinite, chlorite, or fine-grained micas do not exhibit swelling or shrinkage. Vermiculite is intermediate in its swelling and shrinking characteristics.

9. **Dispersion and Flocculation**: As long as the colloidal particles remain negatively charged, they repel each other and the suspension remains stable. If on any account they lose their charge, or if the magnitude of the charge is reduced, the particles coalesce, form loose aggregates and subsequently settle down. This phenomenon of coalescence and formation of flocks is called as *flocculation*. The reverse process of the breaking up of flocks into individual particles is known as *de-flocculation* or *dispersion*.

10. **Brownian Movement**: When a suspension of colloidal particles is examined under a microscope the oscillation of the particles seems visible. The oscillation is attributed to the collision of colloidal particles with those of the liquid in which they are suspended. Suspended soil colloidal particles are always in a state of constant motion. The smaller the particle size, the more rapid is its movement.

11. **Non-permeability**: Colloids are unable to pass through a semi-permeable membrane. Even though the colloidal particles are extremely small, they are bigger than molecules of crystalloids dissolved in water. The membrane allows the passage of water and dissolved substances through its pores, but retains the colloidal particles.

19.2. Types of Soil Colloids

Soil colloids are either inorganic like layered silicate clays, iron and aluminum oxide clays, allophane and associated amorphous clays or organic as humus colloid. Silicate clays are also known as phyllosilicates (*Phyllon* - leaf) because of their leaf-like or plate like structure. These are made up of two kinds of horizontal sheets. One dominated by silicon and other by aluminum and/or magnesium.

Inorganic colloids	Organic colloids
Derived from rocks and parent materials; consist minerals of secondary origin called, "clay minerals"	Derived from decomposition of organic matter; called, "humus"
3 types: Crystalline clay minerals Amorphous clay minerals Primary minerals occurring in clay-size fractions	2 types: Free state humus colloids Clay-humus complex

1. Crystalline Silicate Clay Minerals

Crystalline silicate clay minerals are built of alternating silica tetrahedron and alumina octahedron layers.

Silica tetrahedron: The basic building block for the silica-dominated clays is a unit composed of one central silicon atom surrounded by four oxygen atoms. It is called the *silica tetrahedron* because of its four-sided configuration. An interlocking array or a series of these silica tetrahedra tied together horizontally by shared oxygen anions gives a tetrahedral sheet.

Alumina octahedron: Aluminium and/or magnesium ions are the key cations surrounded by six oxygen atoms or hydroxyl groups giving an eight-sided building block termed as *octahedron*. Numerous octahedra linked together horizontally constitute the octahedral sheet.

An aluminum-dominated sheet is known as *di-octahedral sheet*, whereas the one dominated by magnesium is called as *tri-octahedral sheet*. The distinction is owing to the fact that two aluminum ions in a di-octahedral sheet satisfy the same negative charge from surrounding oxygen and hydroxyls as three magnesium ions in a tri-octahedral sheet.

The tetrahedral and octahedral sheets which are the fundamental structural units of silicate clays are bound together within the crystals by shared oxygen atoms into different layers. The specific nature and combination of sheets in these layers vary from one type of clay to another and control the physical and chemical properties of each clay.

2. Iron and Aluminum Oxide Clays (Sesquioxide Clays)

Under conditions of extensive leaching by rainfall and due to prolonged intensive weathering of minerals in humid warm climates, most of the silica and alumina in primary minerals are dissolved and slowly leached away. The remnant materials, which have very low solubility are the sesquioxides which are mixtures of aluminum hydroxide [$Al(OH)_3$] and iron oxide [Fe_2O_3] or iron hydroxide [$Fe(OH)_3$]. The term sesquioxide was derived from the Latin word *sesqui* which means one and a half, as these oxides had one and one-half times more oxygen than Al and Fe. These clays can grade from amorphous to crystalline. Examples of iron and aluminum oxides common in soils are gibbsite ($Al_2O_3.3H_2O$) and geothite ($Fe_2O_3.H_2O$). Less is known about these clays than about the layered silicates. These clays do not swell, nor are they sticky but have high phosphorus adsorption capacity.

3. Allophane and Other Amorphous Minerals

These silicate clays are mixtures of silica and alumina and are amorphous in nature, with high anion exchange capacity. Even mixture of other weathered oxides (iron oxide) may be a part of the mixture. Typically, these clays occur where large amounts of weathered products exist and are common in soils formed from volcanic ash (e.g., Allophane). Almost all of their charge is from accessible hydroxyl ions (OH^-), which attract positive ions or lose the attached H^+ ions. These clays have a variable charge that depends on H^+ in solution (soil acidity).

4. Humus (Organic Colloid)

The organic constituent of soil is humus which consists of a large group of compounds formed by the decomposition of plant and animal residues. Humus is amorphous, dark brown to black, nearly insoluble in water, but mostly soluble in dilute alkali (NaOH or KOH) solutions. Humus colloids are, amorphous in nature. They are composed basically of carbon, hydrogen, and oxygen, rather than of silicon, aluminum, iron, oxygen and hydroxyl groups. The negative charges of humus are associated with partially dissociated enolic (-OH), carboxyl (-COOH), and phenolic groups.

The important surface phenomenon associated with the colloidal humic compounds is the cation exchange capacity (CEC). The CEC values for organic colloids vary from 250 to 450 meq per 100 g. The organic colloids contribute as much as 20 to 50 per cent of total CEC of soil, even though they constitute only 3 to 5 per cent of the total soil mass.

19.3. Significance of Soil Colloids

The organic and inorganic contaminants are often transported *via* colloidal particles. Majority of surface area and electrostatic charge in a soil resides in the less than 1 μm size fractions, with particles with radii between 20 and 1000 mm constituting the major part of soil surface area. Since the major part of the surface area is due to the colloidal fraction of the soil, almost all surface-controlled processes including adsorption reactions, nucleation and precipitation involve colloids. In addition, colloids are mobile in soils, affecting not only the chemical transport of otherwise immobile chemicals, but also exerting a strong influence on soil hydraulic properties.

20

Clay Minerals

Structure of Crystalline Clays

Soil clay minerals are of two types: crystalline and amorphous. Crystalline clays are made up of sheets of octahedra and tetrahedra of O^{2-} and OH^{-}. Clays with alternating sheets of aluminium octahedra and silicon tetrahedra, are called, "1:1 clays" or "diphormic clays". A sheet of octahedra is sandwiched between two sheets of tetrahedra in 2:1 clays which are called as, "triphormic clays".

The building blocks of clay crystals are called "unit cells". The geometry of octahedra has 3 voids in each half unit cell, equivalent to 6 negative valences. In order to nullify these 6 negative valences, 6 positive valences are needed. The 6 negative valences can be nullified by 3 divalent cations, one each in the three voids, or by 2 trivalent cations, leaving the third void empty. When trivalent cations such as Fe^{3+} and Al^{3+} fill two of the three voids, the clays are called, "dioctahedral clays". When three divalent cations such as Fe^{2+}, Mg^{2+} and Mn^{2+} fill up all the three voids, such clays are called, "trioctahedral clays".

20.1. Layered Silicate Clays

These clays are comprised of two kinds of horizontal sheets, one dominated by silicon and the other by aluminum and/or magnesium. The tetrahedral and octahedral sheets form the fundamental structural units of silicate clays. They are bound together within the crystals by shared oxygen atoms into different layers. The specific nature and combination of sheets in these layers vary from one type of clay to another and largely influence the physical and chemical properties of clays.

Silica Tetrahedron

The basic building block for the silica-dominated sheet is a unit composed of one silicon atom surrounded by four oxygen atoms and it is called *silica tetrahedron* because of its four-sided configuration (Fig. 20.1). An interlocking array or a series of these silica tetrahedra are tied together horizontally by shared oxygen anions giving a tetrahedral sheet.

Alumina Octahedron

Aluminium and/or magnesium ions are the major cations in the second type of sheet. An aluminium (or magnesium) ion is surrounded by six oxygen atoms or hydroxyl group, giving an eight-sided building block called *octahedron*. Numerous octahedra are linked together horizontally to form the octahedral sheet (Fig. 20.1).

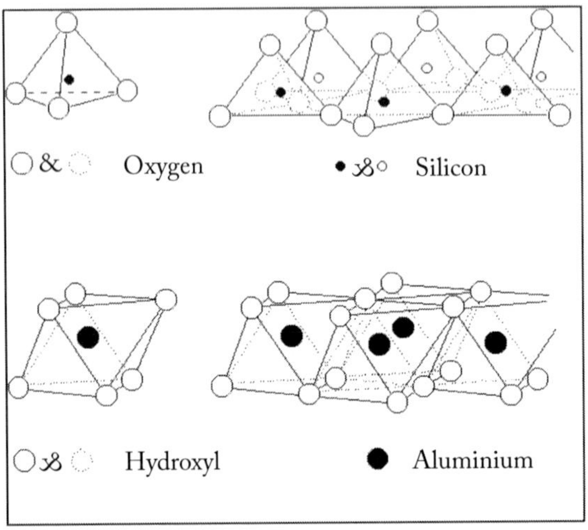

Fig. 20.1: The Si-tetrahedron and Al-octahedron fitting together to form layers

20.2. Classification of Crystalline Clays

Basically, there are 4 types of crystalline clays based on the octahedron and tetrahedron layers:

- Two-layer clays
- Three-layer clays
- Four-layer clays
- Chain-structure clays

1. Two-layer clays (1:1 type)
   a) Equidimensional [Kaolinite group]
   b) Elongate [Halloysite group]
2. Three-layer clays (2:1 type)
   i) Expanding lattice

a) Equidimensional:

[Montmorillonite – Full expansion

Vermiculite – Limited expansion]

b) Elongate: [Nontronite, Saponite, Hectorite]

ii) Non-expanding lattice [Illite group]

3. Four-layer clays (2:1:1 type)

Chlorite group

4. Chain-structure clays

Attapulgite group

Some important characteristics of the main groups of crystalline clays are furnished in Table 20.1.

1. 1:1 Type Minerals

The 1:1-type minerals are made up of one tetrahedral (silica) sheet combined alternately with one octahedral (alumina) sheet and hence the name 1:1-type. In soils, kaolinite is the most prominent member of this group, which includes halloysite, nacrite and dickite as other members. The tetrahedral and octahedral sheets in a layer of a kaolinite crystal are bound together tightly by oxygen atoms, which are mutually shared by the silicon and aluminum cations in their respective sheets. In turn, these layers are held together by hydrogen bonding. Consequently, the structure is fixed and therefore there is no possibility of expansion between layers when the clay is wetted. As water and cations cannot enter between the structural layers of a 1:1-type mineral particle, the effective surface of kaolinite is restricted to its outer faces or to its external surface area. Also, there is little isomorphous substitution in 1: 1- type mineral. This, along with the relatively low surface area of kaolinite, accounts for its low capacity to adsorb cations. Owing to the strong binding forces between their structural layers, kaolinite particles are not readily broken down into extremely thin plates. Kaolinite exhibits very little plasticity, cohesion, shrinkage and swelling.

2. 2:1 Type Minerals

The crystal units of these minerals are characterized by an octahedral sheet being sandwiched between two tetrahedral sheets. Three general groups have this basic crystal structure.

i) Expanding type: Smectite and vermiculite

ii) Non-expanding type: Mica (illite)

Expanding Type

The smectite group is known for interlayer expansion, which occurs by swelling when the minerals are wetted; the water entering the interlayer space forces the layers apart. Montmorillonite is the most prominent member of this group, although beidellite, nontronite, and saponite are also found in soils.

The flake-like crystals of smectites (e.g., Montmorillonite) are composed of an expanding lattice 2:1-type clay mineral. Each layer is made up of an alumina octahedral sheet sandwiched between two tetrahedral silica sheets. There is little attraction between the oxygen atoms in the bottom tetrahedral sheet of one unit and those in the top tetrahedral sheet of another. This permits a readily variable space between layers which is occupied by water and exchangeable cations. This internal surface far exceeds the surface around the outside of the crystal. In montmorillonite, magnesium has replaced aluminum in some sites of the octahedral sheet. In the same way, some silicon atoms in the tetrahedral sheet may be replaced by aluminum. Each of these substitutions gives rise to a negative charge. Therefore, these minerals show high cation exchange capacity and marked swelling and shrinkage properties. Wide cracks commonly form in smectite-dominated soils (e.g., Vertisols) during dry periods. The dry aggregates or clods are very hard, rendering such soils difficult to till.

In Vermiculites, also 2: 1-type minerals, an octahedral sheet occurs between two tetrahedral sheets. In most soils the octahedral sheet in vermiculites is aluminum-dominated (dioctahedral), although magnesium dominated (trioctahedral) vermiculites are also common. In the tetrahedral sheet of most vermiculites, considerable substitution of aluminum for silicon takes place which accounts for most of the very high net negative charge associated with these minerals. Water molecules, magnesium and other ions are strongly adsorbed in the interlayer space of vermiculites. They serve primarily as bridges holding the units together. The degree of swelling is, therefore, considerably less for the vermiculites than for the smectites. For this reason, vermiculites are considered as limited-expansion clay minerals, expanding more than the kaolinites but much less than the smectites. The cation exchange capacity of vermiculites usually exceeds that of all other silicate clays, including montmorillonite and other smectites, because of very high negative charge in the tetrahedral sheet.

Non-expanding Type

Weathered minerals structurally similar to the micas are found in the clay fraction of soils. Fine-grained micas have a 2:1-type crystal. However, the particles are much larger than those of the smectites. In addition, the major source of charge is in the tetrahedral sheet where aluminum atoms occupy about 20% of the silicon sites. This results in a high net negative charge in the tetrahedral sheet, even higher than that found in vermiculites. To satisfy this charge, potassium ions are strongly attracted in the interlayer space and are just the right size to fit into certain spaces in the adjoining tetrahedral sheets. Potassium thus acts as a binding agent, preventing expansion of the crystal. Hence, fine-grained micas are quite non-expansive.

Hydration, cation adsorption, swelling, shrinkage and plasticity are much less intense in fine-grained micas than in smectites. The fine-grained micas exceed kaolinite with respect to these characteristics, may be due in part to the presence of interstratified layers of smectite or vermiculite. In size too fine-grained mica crystals are intermediate between the smectites and kaolinites, their specific surface area varying from 70 to 100 $m^2 g^{-1}$.

Table 20.1: Chief characteristics of the main groups of crystalline clays

	1:1, diphormic clays	2:1, triphormic clays	
		Expanding type	Non-expanding type
Group members	Equidimensional: Kaolinite Elongate: Halloysite	Equidimensional: Montmorillonite sauconite, beidelite, vermiculite Elongate: Nontronite, saponite, hectorite	Equidimensional: Illite, biotite, muscovite
Structure	Rigid lattice	Lattice expanding & contracting with water content	Rigid lattice. Potassium bonds sheets together
Swelling & shrinkage	Very little	Much	Illite little Micas none
SiO_2/R_2O_3	1.9 – 2.1	4.0 – 6.0	2.1 – 3.2
Isomorphic substitution	None	Al for Si; Mg, Fe, Mn for Al	Al for Si
Adsorption	Small	Large	Intermediate
CEC (meq /100g)	5 – 10	80 – 150	Illite 20 – 40 Micas 2- 5
Specific surface	25 – 50 $m^2 g^{-1}$	~ 750 $m^2 g^{-1}$	75 – 125 $m^2 g^{-1}$
Basal spacing (Å)	7.2	Montmorillonite 17 Vermiculite 14	10

3. 2:1:1 Type Minerals

This group is represented by chlorites, commonly present in a variety of soils. Chlorites are ferro-magnesium silicates with some aluminium present. In a typical chlorite clay crystal 2:1 layers, such as in vermiculites, alternate with a magnesium-dominated trioctahedral sheet. Thus, the crystal unit contains two silica tetrahedral sheets and two magnesium-dominated trioctahedral sheets giving rise to 2:1:1 or 2:2-type structure. Particle size and surface area for chlorites are also about the same as for fine-grained micas. There is no water adsorption between the chlorite crystal units, which explains the non-expanding nature of this mineral.

4. Allophane and Other Amorphous Minerals

Typically, these clays are prevalent in soils from volcanic ash (e.g., Allophane). They have high anion exchange capacity. Almost all their charge is from accessible hydroxyl ions (OH^-), which can attract cations. The variable charge of these clays depends on the H^+ concentration in solution.

5. Humus (Organic Colloid)

Humus is dark brown to black, nearly insoluble in water, but mostly soluble in dilute alkali solutions. Humus colloids are not crystalline, but amorphous and are basically composed of carbon, hydrogen, and oxygen, rather than of silicon, aluminum, iron, oxygen and hydroxyl groups. The negative charges of humus are associated with partially dissociated carboxyl and phenolic groups.

21

Ion Exchange

Ion Exchange Defined

By *ion exchange* is meant "the reversible process by which cations and anions are exchanged between solid and liquid phases, and between solid phases, if in close contact with each other". For exchange of cations the term "base exchange" is exclusively used.

21.1. Origin of Ion Exchange Property

A soil owes its ion exchange property almost entirely to the clay and silt fractions (< 20μm) and organic matter. Though soil colloids are amphoteric in nature, they generally carry a net negative charge. The negative charges of soil colloids are due to:

- isomorphous ion substitutions
- ionization of OH- groups attached to Si of broken tetrahedral planes
- Humic (-COOH, -OH), phosphoric and silicic acids

The positive charges of soil colloids are due to:

- Hydrous oxides of Fe, Al, Mn
- Exposed octahedral groups which act as bases by accepting protons from the surrounding soil solution
- Basic groups of humus (due to N)

21.2. Cation Exchange Capacity

Cation exchange capacity (CEC) is defined as "the amount of exchangeable cations per unit weight of dry soil". It is expressed in milliequivalents of cations per 100 gram of soil (meq/100g) or cmole (+) kg^{-1} of soil. Cation exchange takes place on surfaces of clay and humus colloids. The higher the CEC of soil the greater is the amount of cations it can retain.

21.2.1. Mechanism of Cation Exchange

The exchange of cations can be explained on the basis of the electro-kinetic theory. By this theory, the adsorbed cations constituting the outer shell of the ionic double layer are assumed to be in a state of oscillation when suspended in water, forming a diffuse double layer. During these oscillations, some of the cations move away from the surface of the clay micelle. In the presence of an electrolyte solution, a cation of the added electrolyte slips in between the inner negative layer and the outer oscillating positive ion. The electrolyte cation is now adsorbed on the micelle and the surface cation remains in solution as an exchanged ion. Thus, the exchange of cations takes place.

Cations are positively charged nutrient ions and molecules, while clay particles are negatively charged constituents of soils. These negatively charged clay particles attract, hold and release positively charged nutrient cations. Organic colloids also have negative charges to attract cations. Sand particles, however, carry little or no charge and do not react. Cations held by soils can be replaced by other cations which means they are exchangeable. For example, Ca^{++} can be exchanged for H^{+} or K^{+} and *vice versa* (Fig. 21.1).

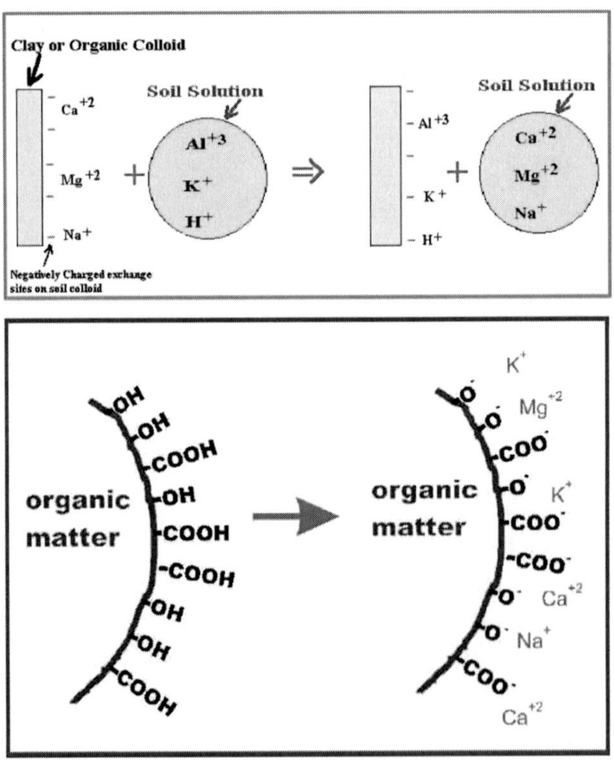

Fig. 21.1: Schematic depiction of CEC

The CEC depends on the amount and kind of clay and organic matter present. A high-clay soil holds more exchangeable cations than would a low-clay soil and retains large amounts of cations against potential leaching losses. Sandy soils, on the other hand, retain smaller quantities due to low CEC. The CEC also increases with organic matter content. Clay minerals usually have CEC values ranging from 10 to 150 meq/100 g. The CEC of organic matter varies from 200 to 400 meq/100 g.

21.2.2. Replacing Power of Cations

The replacing power of cations in the soil exchange complex varies with the type of ion, its valence, size, concentration, degree of hydration and the kind of clay mineral involved. As it is controlled by several factors no single order of replacement can be given. When all other factors are equal, the replacing power of monovalent cations increases in the following order: Li < Na < K < Rb < Cs < H and for divalent cations: Mg < Ca < Sr < Ba. In a mixture of monovalent and divalent cations, as they exist in normal soils, the replacing power increases in the following order: Na < K < NH_4 < Mg < Ca < H. This implies Na is more easily replaced than K and K more easily than NH_4, and so on.

21.2.3. Percent Base Saturation

Percent base saturation is the percentage of the total cation exchange capacity occupied by the basic cations such as Ca^{2+}, Mg^{2+} and K^+. Basic cations are distinguished from the acid cations H^+ and Al^{3+}.

The percentage base saturation is expressed as follows:

$$\%BS = [(Ca^{2+} + Mg^{2+} + K^+)/CEC] \times 100$$

The soils with a high percent of base saturation are considered fertile because:

- They have little or no acid cation Al^{3+} that is toxic to plant growth.
- Soils with high base saturation have a higher pH; therefore, they are more buffered against acid cations from plant roots and soil processes (nitrification, acid rain, etc.) that acidify the soil.
- They contain greater amounts of the essential plant nutrient cations K^+, Ca^{2+} and Mg^{2+} for crop uptake.

To calculate the percent base saturation, divide the sum of the K, Mg, Ca, and Na (the bases) in meq/100g soil by the total CEC and multiply the result by 100.

Example: With the following data, calculate the per cent BS saturation of a given soil:

- K = 0.32 meq/100g soil
- Mg = 0.16 meq/100g soil
- Ca = 1.20 meq/100g soil
- Na = 0.02 meq/100g soil
- Al = 0.20 meq/100 g soil
- CEC = 4.25 meq/100g soil

Total for bases = K + Mg + Ca + Na = 1.70 meq/100g soil

Per cent base saturation = (1.70 ÷ 4.25) x (100) = 40%

21.2.4. Importance of Cation Exchange

Cation exchange is the key to soil productivity in causing and correcting soil acidity and basicity, in altering soil physical properties and as a mechanism in purifying percolating waters. Most plant nutrients are available to crops only from the soil exchange complex. Cation exchange is very important in soils because of the following considerations:

- The exchangeable K is a major source of plant K.
- The exchangeable Mg is often a major source of plant Mg.
- The amount of lime required to raise the pH of an acidic soil is greater as the CEC is greater.
- Cation exchange sites hold Ca^{2+}, Mg^{2+}, K^+, Na^+, and NH_4^+ ions and slow down their losses by leaching.
- Cation exchange sites hold fertilizer K^+ and NH_4^+ and greatly reduce their mobility.
- Cation exchange sites adsorb many metals (Cd^{2+}, Zn^{2+}, Ni^{2+}, and Pb^{2+}) that might be present in wastewater, thereby cleansing the water that drains into groundwater.
- The timing and application rates are important in planning a fertilizer programme. For example, it may not be wise to apply K on very sandy soils in the middle of a monsoon, where rainfall can be high and intense. Fertilizer application should be split to prevent leaching losses. Splitting N applications helps meet peak crop demand and reduces the potential for nitrate leaching in soils.

21.3. Anion Exchange Capacity

The sum total of exchangeable anions that a soil can adsorb is known as anion exchange capacity. It is expressed as c mol kg^{-1} or meq /100 g soil. The intensity of adsorption and exchange of anions is related to the type of clay mineral, soil reaction and the nature of anion.

Kaolinitic minerals have a greater anion adsorbing and exchange capacity than montmorillonite and illite clays. The exchange capacity for anions is less in the case of montmorillonite and illite because the exchange is located at only a few broken bonds.

The lower the pH the greater is the adsorption of anions. All anions are not adsorbed equally readily. Some anions such as H_2PO_4 are adsorbed very readily at all pH values in the acid as well as in the alkaline range. The affinity for adsorption of the anions commonly present in soil follows the order: $NO_3 < Cl < SO_4 < PO_4$.

Importance of Anion Exchange

Anion exchange assumes importance particularly in relation to the fixation phosphate ions. The exchange is achieved mainly by the replacement of OH^- ions of the clay mineral. The reaction is very similar to cation exchange:

$$OH - Clay + H_2PO_4 \rightarrow H_2PO_4 - Clay + OH^-$$

Adsorption of phosphate ions by clay particles from soil solution limits its availability to plants and this is known as *phosphate fixation*. Because of the reversible nature of the reaction, the phosphate ions may again become available when they are replaced by OH^- ions of liming materials applied to soil to correct soil acidity. Hence phosphate fixation is temporary. But the whole of the phosphate adsorbed by clay is not exchangeable, even at pH 7.0 and above. So, substantial quantities of phosphate ions may still be retained by clay particles. The OH^- ions originate also from hydrous oxides of iron and aluminium in the soil. Therefore, the phosphate ions react with the hydrous oxides also, form insoluble hydroxy phosphates of iron and aluminium and get fixed as in the case of silicate clays.

$$\underset{\text{(Soluble)}}{Al(OH)_3} + H_2PO_4 \rightarrow \underset{\text{(Insoluble)}}{Al(OH)_2 . H_2PO_4} + OH^-$$

If the reaction takes place in slight acidity conditions, it is reversible and soluble phosphate is again liberated when hydroxy phosphate comes in contact with ions. In strongly acid conditions, the phosphate ions are irreversibly fixed and are totally unavailable for the plants.

22

Soil Reaction

Soil Reaction

Soil reaction is an indication of the acidity or basicity of the soil and is measured in pH units. The pH scale ranges from 0 to 14, with 7.0 as neutral point. From pH 7 to 0, soil is increasingly acidic. From pH 7 to 14, soil is increasingly alkaline.

Pure H_2O ionizes to a very small degree, though to a measurable extent, so that it has the following equilibrium: both H^+ and OH^- are in concentrations of 10^{-7} moles L^{-1}.

$$HOH \leftrightarrow H^+ + OH^-$$

According to the Law of Mass Action,

$$\frac{[\text{Concentration of } H^+ \text{ ions}] \times [\text{Concentration of } OH^- \text{ ions}]}{[\text{Concentration of undissociated water}]} = \text{Constant}$$

$$\text{Or, } \frac{[H^+] \times [OH^-]}{[HOH]} = \text{Constant}$$

Since the amount of undissociated water is extremely large (only one out of nearly 550 water molecules dissociates which is so close to unity that it can be taken as unity), the above equation can be rewritten as:

$$[H^+] = [OH^-] = K_w$$

This ionic product (K_w) of water has been measured by electrical conductivity measurements to be 1×10^{-14} gram ions per litre at 24 °C. In pure water, the number of H^+ ions is equal to the number of OH^- ions and so the concentration of each must be 1×10^{-7} gram ions per litre. If the concentration of $[H^+]$ ions is more than 10^{-7}, the solution is acidic and *vice versa.*

22.1. pH Defined

Numbers such as 1.0×10^{-7} or 0.0000001 are not convenient to use. Hence, Sorensen in 1909 suggested that the H^+ concentration be generally expressed as the negative power to which 10 must be raised in order to express the required concentration and this value be designated by the symbol pH.

The term "pH" is derived from the French *pouvoir hydrogene* which means "hydrogen power". pH is defined as "*the negative logarithm of the hydrogen ion [H⁺] concentration in solution*".

i.e., $[H^+] = 10^{-pH}$

$\log [H^+] = - pH \log 10$

$$\text{or, } pH = \frac{-\log [H^+]}{\log 10}$$

Since log 10 = 1.0,

$pH = -\log [H^+] = \log \{1 \div [H^+]\}$

Doubling the $[H^+]$ concentration lowers the pH by 0.3 units at any point along the pH scale. The $[H^+]$ has 10-fold change between each whole number: Thus, a soil of pH 5 has 100 times more H^+ in solution than a soil with pH 7. The OH^- concentration can be measured on a pOH scale just as H^+ is measured by pH. The sum of pH and pOH is always 14. When the pH goes up, pOH goes down by a corresponding magnitude and *vice versa.* When H^+ is 10^{-4}, OH^- is 10^{-10} or, when pH is 4, pOH is 10 (Table 22.1).

Table 22.1: Relationship between acidity and alkalinity

pH	Acidity $[H^+]$ in g ions/Litre	Alkalinity $[OH^-]$ in g ions / Litre	pOH
0	1.0 (10^0)	0.000 000 000 000 01	14
1	0.1 (10^{-1})	0.000 000 000 000 1	13
2	0.01 (10^{-2})	0.000 000 000 001	12
3	0.001 (10^{-3})	0.000 000 000 01	11
4	0.0001 (10^{-4})	0.000 000 0001	10
↓	↓	↓	↓
11	0.000 000 000 01	0.001	3
12	0.000 000 000 001	0.01	2
13	0.000 000 000 000 1	0.1	1
14	0.000 000 000 000 01	1.0	0

22.2. Soil pH

Most soils have pH between 4 and 8. Nearly all soils with pH values > 8.5 have a high percentage of Na^+ ions on their cation exchange sites. Soils with pH values < 4 generally contain sulphuric acid. An evaluation of pH in description terms is given Fig. 22.1. below:

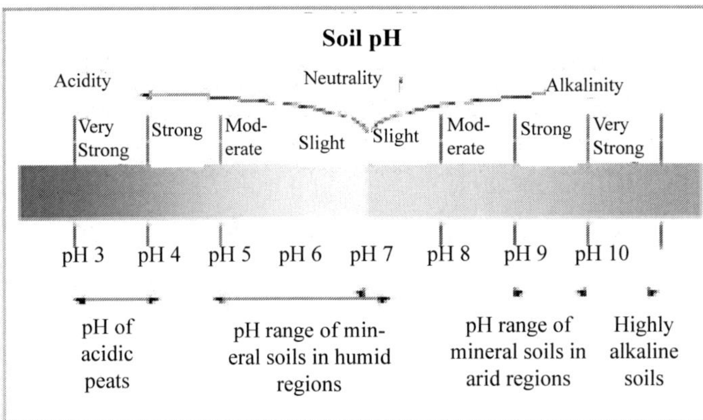

Fig. 22.1: Description of soil pH

22.3. Significance of pH

The pH can be easily measured and can indicate much about a soil. For instance, a strongly acidic soil indicates that it is intensively leached; low in base saturation and microbial activity; and is mostly kaolinite clay. The pH influences plant growth, as the availability of all plant nutrients is influenced by pH (Fig. 22.2). Most N-fixing bacteria and bacteria that act on organic matter are not active in strongly acidic soils. Fungi usually tolerate acidity better. Soils high in Ca^{++} may have pH close to 8.5. With increased Na^+, pH may be > 8.5 – 10.0.

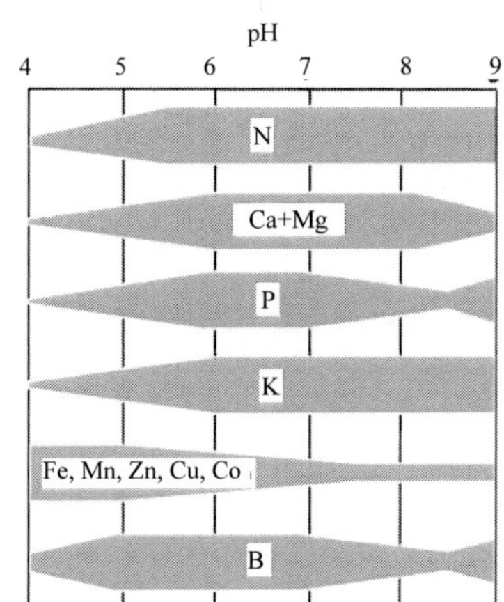

Fig. 22.2. Availability of plant nutrients in relation to soil pH

22.4. Kinds of Acidity

Active Acidity refers to the H^+ ion concentration in soil solution. It is expressed as pH value.

Reserve acidity or *Exchange acidity* or *Potential acidity* refers to the amount of acidity that must be neutralized to bring an acid soil to neutrality. This acidity is due the H^+ and an Al^{3+}ions held on the soil colloids and is expressed in milli equivalents / unit soil mass. Reserve acidity may be 1,000 times greater than the active acidity in a sandy soil and 50,000 to 100,000 times greater in clay soil.

22.5. Buffering in Soils

Most soils can resist appreciable changes in pH when large amounts of either strongly acidic or basic materials are added. The ability of the soil to resist a change in pH is called its "*buffering capacity*". The buffering capacity increases as the CEC increases. Soils with high contents of clay and organic matter have high buffering capacity as compared to sandy soils with low amounts of organic matter. To exhibit buffering, a soil must "remove" H^+ ions of added acids or neutralize the hydroxyls (OH^-) of added bases. This occurs by cation exchange and neutralization:

$$\text{Colloid-H} + NH_4OH \longleftrightarrow \text{Colloid-}NH_4 + HOH$$

$$\text{Colloid-2H} + CaCO_3 \longleftrightarrow \text{Colloid-Ca} + CO_2 + HOH$$

$$\text{Colloid-Ca} + 2H_2CO_3 \longleftrightarrow \text{Colloid-H} + Ca(HCO_3)_2$$

22.6. Factors Affecting Soil pH

Soil pH depends on a variety of factors such as (a) five soil forming factors, (b) season of the year, (c) cropping practices, (d) soil horizon sampled, (e) water content at sampling time and (f) the way the pH is determined.

Vegetation influences soil pH in a complex way because it produces organic matter and influences leaching. The addition of decomposable organic matter to soil results in the formation of organic acids. These organic acids add to the cation exchange capacity, but the % base saturation and pH are lowered. Leaching processes are accelerated under acid conditions because more cations are released by acid weathering and fewer are held by cation exchange. In temperate regions, soils formed under grasses are usually less acidic than those formed under trees. The reason is grasses utilize more bases and therefore deposit more bases on the soil surface than do the trees. Many coniferous trees are light feeders on bases and tend to produce more acidic soils.

22.7. Measurement of pH

The pH measurements are made in the laboratory by the accurate electrometric methods (pH meters). In the field, pH is measured by portable pH meters or roughly by colour indicator papers.

23

Soil Electrical Conductivity

Electrical conductivity (EC) is the ability of a material to transmit electrical current. *In the soil, the EC is the ability of the soil solution to carry an electrical current and is a measure of the total amount of soluble salts present in the soil.* Pure water is a poor conductor of electricity and increases in soluble salts water result in proportional increases in the conductivity.

23.1. Sources of Soluble Salts in Soil

The soluble salts in soils consist mostly of the cations sodium, calcium, and magnesium and the anions chloride and sulfate. Constituents that ordinarily occur only in minor amounts are the cation potassium and the anions bicarbonate, carbonate, and nitrate. The original and, to some extent, the direct source of all the salt constituents are the primary minerals found in soils and in the exposed rocks of the earth's crust.

While the mean chlorine and sulfur content of the earth's crust is 0.05 and 0.06 percent, respectively, sodium, calcium, and magnesium each occur in the range of 2 or 3 percent. During the process of chemical weathering, which involves hydrolysis, hydration, solution, oxidation, and carbonation, these constituents are gradually released and made soluble. When carbon-di-oxide, of atmospheric or biological origin, dissolves in water, bicarbonate ions are formed. Water containing carbon-di-oxide is an active chemical weathering agent that releases appreciable quantities of the cation constituents as the bicarbonates. The concentrations of carbonate and bicarbonate ions are a function of the pH of the solution. Appreciable amounts of carbonate ions can be present only at pH values of 9.5 or higher.

There are soils where high concentrations of nitrate are found. Boron which is toxic to plants even in low concentrations and whose principal source is the mineral tourmaline, a constituent of primary rocks, also deserves mention.

Although weathering of primary minerals is a source of nearly all soluble salts, saline soils usually occur in areas that receive salts from other locations, with water being the primary carrier. The ocean may be the source of salts in soils where the parent material consists of marine deposits buried during earlier geologic periods and have since been uplifted. The ocean is also the source of the salts in low-lying soils along seacoasts. Sometimes salt is transported inland as spray by winds and is called, "cyclic salt". However, more commonly, the direct source of salts is surface and ground waters all of which contain dissolved salts, the concentration being related to the salt content of the soil and geologic sources with which the water has been in contact. Irrigation waters act as sources of salts and may add salts to soils under natural conditions, as when they flood low-lying land or when ground water rises close to the soil surface.

23.2. Significance of Electrical Conductivity

The EC of soil water is a good indication of the amount of nutrients available for crops as either cations or anions. These ions dissolved in the soil water carry electrical charge and thus determine the EC level of soil. Where EC levels are high (more available nutrients) less fertilizer is needed. The EC is harmoniously related to soil properties that affect crop productivity, including soil texture (sands have a low conductivity, silts have a medium conductivity, and clays have a high conductivity), cation exchange capacity, drainage conditions, organic matter level and salinity. Thus, knowing a soil's EC, one can predict plant health in that soil.

23.3. Testing the Electrical Conductivity

The EC is measured using an EC meter. The probe consists of two metal electrodes and a constant voltage is applied across the electrode, resulting in an electrical current flowing through the sample. The current flowing through water is proportional to the concentration of dissolved ions in the water.

The unit of measurement for EC is deci-Siemens per metre (dS m^{-1}). (1 dSm^{-1} = 1milli mhos cm^{-1}). The unit "mhos" derives from the standard name for electrical resistance reflecting the inverse relationship between resistance and conductivity. Since the electrical current flow increases with increasing temperature, the EC values are to be corrected to a standard temperature of 25 °C. A good EC meter will have automatic temperature compensation so as to get accurate results regardless of sample temperature.

A good soil EC level will be <1.0 dS m^{-1}. The critical range is 1.0 to 3.0 dS m^{-1}. A soil whose EC is in excess of 4 dS m^{-1}, but the exchangeable sodium percentage is < 15, is considered as a saline soil. The relationship between EC and total soluble salts (TSS) in soil is:

TSS (in mg $litre^{-1}$ or ppm) = EC (in mmhos cm^{-1} or dS m^{-1}) x 640.

24

Soil Organic Matter

Soil organic matter (SOM) is one of the most complex materials existing in nature. Besides the organic constituents in undecayed plant and animal tissues, SOM contains both living and dead microbial cells, microbially synthesized compounds and an unending array of derivatives of these materials produced as the result of microbial activity. Soil organic matter contains most of the naturally occurring organic compounds. Some components of SOM are undoubtedly unique to the soil environment, particularly those involving inorganic-organic complexes.

24.1. Composition of Soil Organic Matter

The composition of soil organic fraction is as follows:

1. Components of organic residues undergoing decomposition
2. Metabolic products of micro-organisms utilizing organic residues as a source of energy
3. Products of secondary synthesis in the form of bacterial plasma
4. Humic substances

The first three categories consist of various non-specific substances such as proteins, decomposition products of proteins, carbohydrates, organic acids, fats, resins, waxes, etc., and constitute 10 – 15% of soil organic matter. Humic substances constitute 85 – 90%.

Soil Organic Matter

- Humus
  - Strictly humic substances
    Humic acid
    Fulvic acid
    Humin
    Hymatomelanic acid
  - Products of advanced decompositions of organic residues Products of resynthesis by microbes [Proteins, carbohydrates, fats, waxes, lignins, tannins, etc.]
- Fresh & incompletely decomposed plant & animal residues

24.2. Proximate Constituents of Soil Organic Matter

Included in the extremely diverse group of compounds that constitute SOM are:

a. Carbohydrates and related compounds [mono- and disaccharides, celluloses, hemicelluloses, pectins, pentosans, mannans, polyuronides, uronic acids, organic acids, alcohols, hydrocarbons, aromatics]

b. Proteins and their derivatives [amino acids, amides, amino sugars, nucleoproteins, purine and pyrimidine bases]

c. Lignins and their derivatives

d. Fats and related substances

e. Tannins and their derivatives

f. Resins and terpenes

The composition of plant residues that constitute soil organic matter is furnished in Table 24.1.

Table 24.1: The composition of proximate constituents of plant residues

S. No.	Constituent	Composition (%)
1.	Carbohydrates	
	i. Sugars, Starches	1 - 5
	ii. Celluloses	20 - 50
	iii Hemicelluloses	10 - 20
2.	Fats, waxes and tannins	1 - 8
3.	Proteins (simple & crude)	1 - 15

24.3. Decomposition of Soil Organic Matter

The organic fraction of the soil is a complex assemblage of substances whose composition is determined partly by the plant and animal residues added to the soil, and to a greater extent, by the transformations of these substrates through physical, chemical and biological means. As a result, a multitude of compounds is present: the components of organic residues undergoing decomposition and the metabolic by-products of the microorganisms utilizing such compounds as a source of energy.

Thus, in soil we find two major decomposition processes taking place simultaneously:

1. Breakdown of organic substrates, both fresh and native
2. Synthesis of humic substances

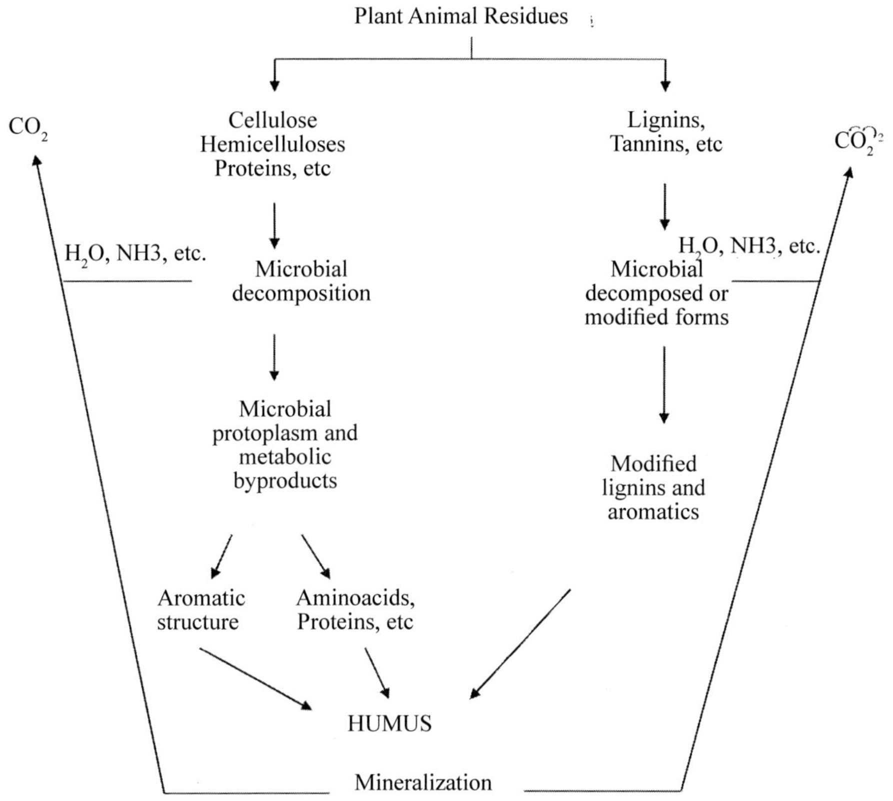

Fig.24.1: Organic matter decomposition and formation of humic substances in soil

25

The Carbon Cycle

The Carbon Cycle

The element that serves as a corner stone for cell structure is carbon. Plant tissues and microbial cells contain approximately 40 to 50% of C on dry weight basis. Yet, the ultimate source is the CO_2 of the atmosphere. The CO_2 is converted into organic C by the action of photoautotrophic organisms (higher plants and green algae). It has been estimated that the vegetation on earth's surface consume some 1.3 x 10^{14} kg of CO_2 per annum which is about $^1/_{20}$ of the total CO_2 supply of the atmosphere or $^1/_{1000}$ of that in the ocean. Soil carbon constitutes the largest terrestrial pool of carbon.

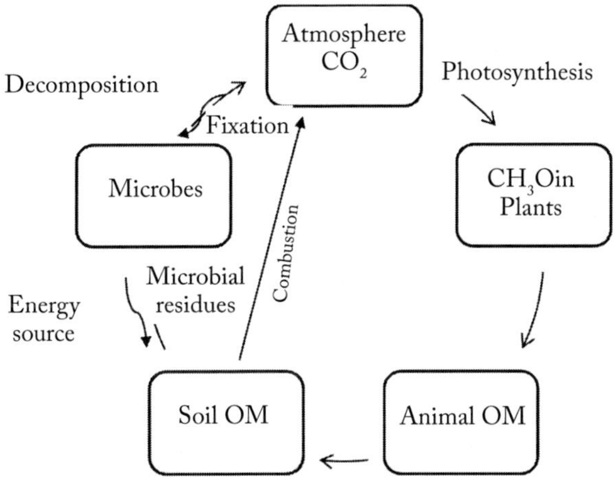

Fig. 25.1: The carbon cycle

25.1. The C: N Ratio

The C: N ratio in the furrow-slice of arable soils will be about 10: 1 to 12:1. The ratio is lower in soils of arid regions than those of humid regions. Also, the ratio is narrower for subsoils, in general, than for the corresponding surface layers. The C: N ratio in the plant residues that are added to the soil is variable, ranging from about 30: 1 for legumes and farmyard manure to as high as 100:1 in some straw residues. The C: N ratio for microbes is much narrower, between 4:1 and 9:1.

25.2. Significance of C: N Ratio

The two most significant factors of C:N ratios is: (1) competition for available N results when residues having a high C:N ratio are added to the soil; (2) due to the constancy of this ratio in soils, the maintenance of carbon (and hence organic matter) is dependent on the soil N level.

Suppose we consider a normal cultivated soil in a condition favouring ammonification and nitrification. The NH_4^+ and NO_3^- are present in large amounts and the C: N ratio is narrow. Now, if a large quantity of organic residues with wide C: N ratio are added to this soil. A change quickly occurs. The heterotrophic flora (bacteria, fungi and actinomycetes) become active and multiply rapidly, yielding CO_2 in large quantities. Under these conditions, NO_3^- practically disappears from the soil due to the insistent microbial demand for this element to build their tissues. And for the time being, little or no N, even NH_4^+, is available to higher plants. As decay occurs, the C: N ratio of the plant material decreases since C is being lost continuously and N is conserved simultaneously.

This condition persists until humification nears completion at which time, the activities of the decay organisms gradually decline due to lack of easily oxidizable C. Their numbers decrease, CO_2 formation drops off, N ceases to be at a premium and nitrification can proceed. Nitrates again appear in quantity and the original conditions again prevail, except, for the time being somewhat richer in both N and hums. The sequence of events, an important phase of C cycle is represented in Fig. 25.2.

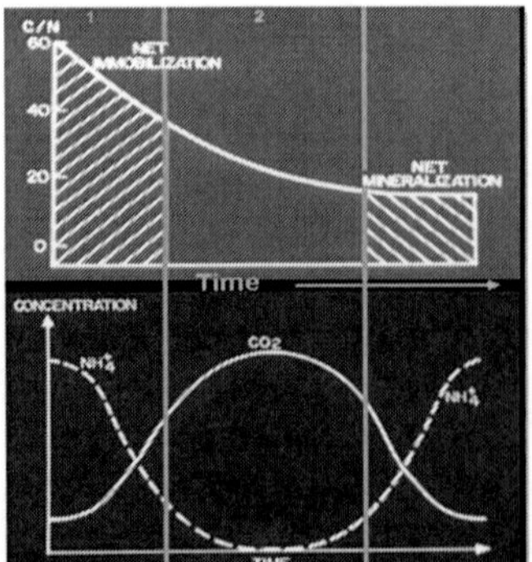

Fig. 25.2: The changes that take place when organic material with high C: N ratio is added to soil

The graphs above show what happens when organic material with high C:N ratio (such as rice straw) is added to soil. The graphs have been divided into three time periods:

1. Initial phase - Net immobilization
   - Number of microorganisms are increasing.
   - Microorganisms consume N (NH_4^+), causing it to decrease.
   - Microorganisms consume C products and release CO_2.
   - C:N ratio is decreasing.
2. Intermediate phase
   - Microorganisms have consumed most of the available N.
   - Maximum consumption of C products and production of CO_2.
3. Final phase – net mineralization
   - Microorganisms are dying.
   - NH_4^+ level is increasing - NH_4^+ that was part of microorganisms is now available in soil.
   - C compounds are decomposed - CO_2 production is low.
   - C:N ratio is low.

25.3. Biomass Carbon

The CO_2 removed from the atmosphere during photosynthesis is converted into organic C and stored in biomass, such as trees and crops. When harvested and combusted, the carbon in the biomass is released as CO_2, thus completing the C cycle. Thus, the carbon that is stored in the biomass is called, *'biomass carbon'*.

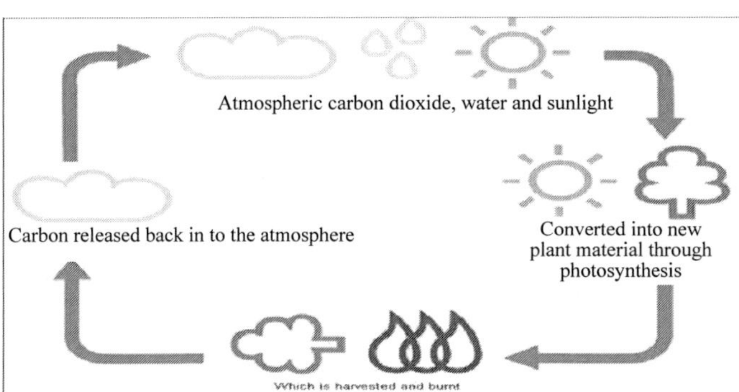

Fig. 25.3: The biomass carbon cycle

The CO_2 released back to the atmosphere is the same CO_2 that was just recently removed or "sequestered" from it. The C in biomass will return to the atmosphere regardless of whether it is burnt for energy or allowed to biodegrade (Fig. 25.3).

26

Humus Formation

The synthesis of humic substances is the least understood and the most intriguing aspect of humus chemistry. Studies on this subject are of enduring and continued research. Several pathways have been identified for the formation of humic substances during the decay of plant and animal remains in soil, the main ones being shown in Fig. 26.1.

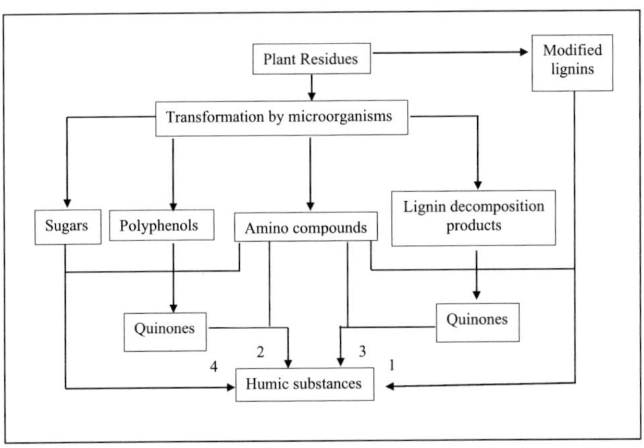

Fig. 26.1: Mechanisms for the formation of soil humic substances

The classical theory, popularized by Waksman, is that humic substances represent modified lignins (Pathway 1) but the majority of present-day investigators favour a mechanism involving quinones (Pathways 2 and 3). In practical terms, all the four pathways must be considered as likely mechanisms for the synthesis of humic and fulvic acids in nature, including sugar-amine condensation (Pathway 4). These four pathways may operate in all soils, but not to the same extent or not necessarily in the same order of importance. Lignin pathway may dictate in poorly drained soils and wet sediments (swamps, etc.), while synthesis from polyphenols may be of considerable significance in certain forest soils. Repeated and subtle fluctuations in temperature, moisture and irradiation in surface soils enveloped in a sharp continental climate may lead to humus synthesis by sugar-amine condensation.

26.1. The Lignin Theory

For long it was thought that humic substances were derived from lignin (Pathway 1). According to this theory, lignin is incompletely utilized by microorganism and the residuum becomes incorporated into soil humus. Lignin modifications include loss of methoxyl (OCH_3) groups with the generation of o-hydroxyphenols and oxidation of aliphatic side chains to form COOH groups. The modified material is prone to further unknown changes to yield first humic acids and then fulvic acids. This pathway, illustrated in Fig. 26.2, is exemplified by Waksman's lignin-protein theory.

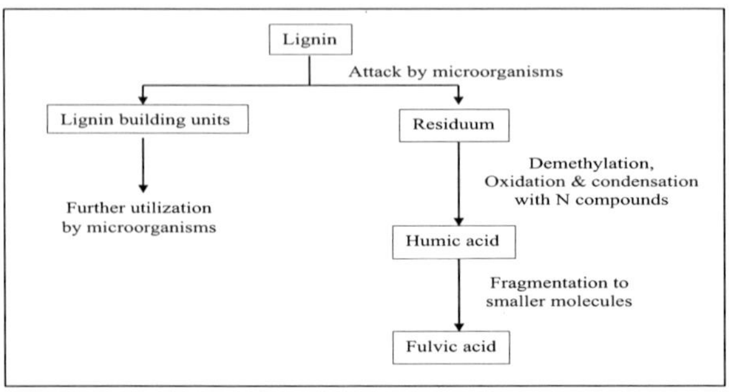

Fig. 26.2: Schematic representation of lignin theory of humus formation

26.2. The Polyphenol Theory

In pathway 3, lignin still plays a crucial role in humus synthesis, though in a different way. Here, phenolic aldehydes and acids released during microbiological attack on lignin undergo enzymatic conversion to quinones, which polymerize to form humic like macromolecules. Pathway 2 is somewhat similar to pathway 3, except that the polyphenols are synthesized by microorganisms from non-lignin C sources [e.g. cellulose]. The polyphenols are then enzymatically oxidized to quinones and further converted to humic substances.

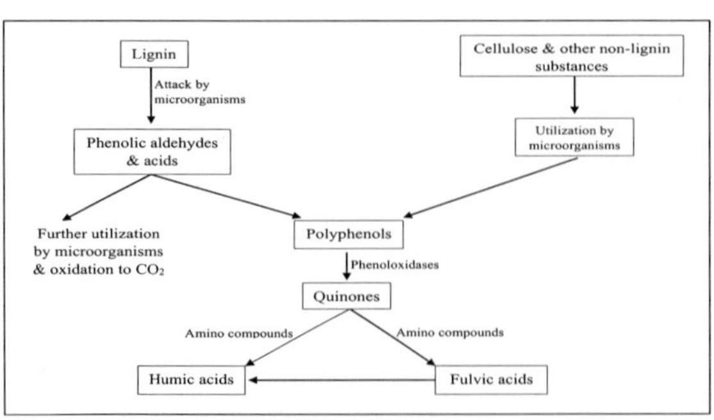

Fig. 26.3: Schematic representation of polyphenol theory of humus

26.3. Sugar-amine Condensation Theory

The belief that humus is formed from sugars dates back to the primitive days of humus chemistry. By this theory, reducing sugars and amino acids formed as by-products of microbial metabolism undergo non-enzymatic polymerization leading to the formation of brown nitrogenous polymers. An important feature of the theory is that the reactants (sugars, amino acids, etc.) are produced in abundance through the activities of microorganisms.

The inceptive reaction in sugar-amine condensation entails the addition of amine to the aldehyde group of the sugar to form the N-substituted glycosylamine which later forms the N-substituted-1-amino-deoxy-2-ketose. This is subjected to (1) fragmentation, with the liberation of the amine and formation of 3-carbon chain aldehydes and ketones (such as acetol, glyceraldehyde and dihydroxyacetone), (2) loss of two water molecules to form reductones, and (3) loss of three water molecules to yield hydroxymethyl furfural. All of these compounds are hyperactive and readily polymerize in the presence of amino compounds to form brown-colored products of unknown composition.

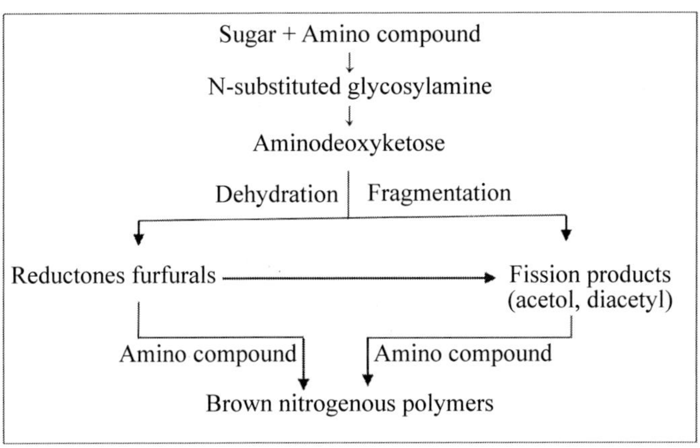

Fig. 26.4: Schematic representation of sugar-amine condensation theory of humus formation

26.4. Humic Substances

Humus is a dark-coloured (brown or dark-brown) complex mixture of amorphous and colloiodal substances arising from modified plant materials and synthesized microbial tissues.

Humic acids: The fraction of humic substances that is not soluble in water under acidic conditions ($pH < 2$) but is soluble at higher pH values is humic acid. Humic acids are the major extractable components of soil humic substances. They are dark brown to black in color (Fig. 26.5).

Fulvic acids: The fraction of humic substances that is soluble in water under all pH conditions is fulvic acid. The fulvic acids remain in solution after removal of humic acid by acidification. Fulvic acids are light yellow to yellow-brown in color.

Humin: The fraction of humic substances that is not soluble in water at any pH value and in alkali is humin. Humin is black in color.

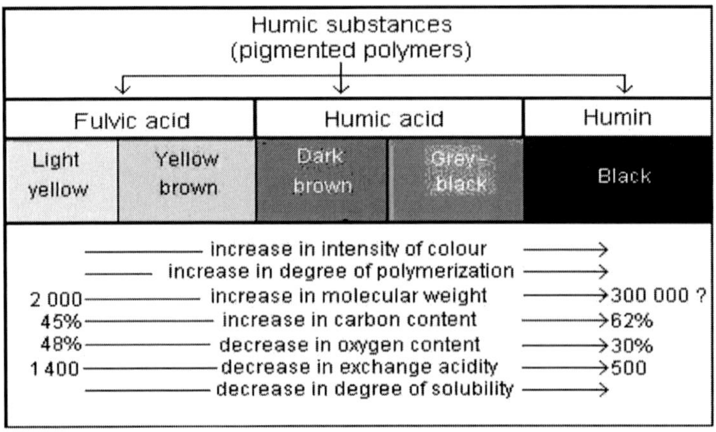

Fig. 26.5: Properties of humic substances

27

Soil Organisms

To millions of different organisms around the world, soil is the habitat. A handful of soil can contain billions of soil micro-organisms. One gram of a fertile soil can harbour up to one billion bacteria. Only 5% of what is produced by green plants is consumed by animals, but 95% is consumed by microorganisms. Soil organisms are generally grouped into microfauna/microorganisms and macrofauna. The main soil microorganisms are bacteria, fungi, and protozoa. The macro-fauna includes oligochaeta, arthropods, mollusks and nematodes. These organisms play an essential role in soil formation and soil environment.

27.1. Bacteria

Bacteria are the major single-celled microorganisms in the soil ecosystem that are responsible for many important processes. There are many different types of bacteria and most of them are aerobic, requiring oxygen from the soil atmosphere. However, there are bacteria that live without oxygen and other types can live both with and without oxygen. The growth of these bacteria is limited by the food that is available in the soil. They decompose plant materials as well as other organisms' wastes that contain plant nutrients in an unusable form and convert them into a form that can be used by plants. One such essential nutrient that is converted by bacteria is nitrogen. Nitrogen-fixing bacteria are able to use the nitrogen (N_2) that exists in the atmosphere and convert it through cellular processes into ammonia (NH_3). Nitrifying bacteria mix ammonia with water to create ammonium ion (NH_4^+) and nitrate (NO_3^-), which can be used by other organisms.

Protozoa are also present in the soil environment and they distribute ammonium into the soil for use by plants and other organisms. They feed on bacteria but do not need as much N as the bacteria provide and, therefore, release the excess ammonium. Denitrifying bacteria return the nitrogen back into the atmosphere as N_2.

27.2. Actinomycetes

Actinomycetes are a type of bacteria that are considered as fungi because they grow hyphae like fungi and they can live in harsh environments in which other bacteria usually do not thrive. They since were renamed as *actinobacteria.* Actinomycetes are capable of decomposing chitin, cellulose, and such other materials in dead plants, animals and fungi, while other organisms cannot disintegrate these materials because of their complex structure. They reduce the chitin, for example, from a polymer into simple sugars and ammonia that other organisms can then use. We appreciate the earthy aroma after a rain shower which actually indicates a healthy environment. That smell - known as "petrichor" - stems from microscopic streptomycete bacteria in the soil that produce a compound called "geosmin".

27.3. Fungi

Soil fungi also constitute a large component of the soil and come in various sizes, shapes, and colors. They are more abundant where the bacteria cannot thrive in environments such as acidic, have low amounts of nitrogen, have low moisture and high complex carbohydrates (bacteria cannot break them down). Fungi are capable of decomposing tough organic matter such as chitin, cellulose, keratin and lignins. Most organisms cannot decompose these materials, but fungi, like actinobacteria, can produce enzymes to degrade those strong compounds into simple sugars that can be digested. Not only are they decomposers, but fungi also are responsible for supplying nutrients to plants. Fungi form a mutualistic relationship with plants called *mycorrhiza*, which allows both parties to benefit. The fungi grow on or within (endopyhtic) the roots of plants. Because of the hyphae's high surface area to volume ratio, fungi gain better access to essential nutrients and deliver them to the plants. The plants, in turn, give the fungi food in the form of glucose. Mushrooms have underground roots (mycelium) that absorb nutrients and water until they are ready to flower in the mushroom form. They tolerate acidity, which makes them very important to decompose materials in very acidic forests that microbes cannot do and they can also decompose lignin, which is the woody tissues for decomposing plants.

27.4. Macro-fauna

Microorganisms, in addition to converting nutrients into usable forms, are themselves the food source for many of the macro-fauna in the soil. The macro-fauna depends on the nutrients that the microbes produce.

The most beneficial macro-fauna in soil is the earthworm. Earthworms are "ecosystem engineers" impacting soil properties as well as nutrient cycles. They consume organic materials and excrete castings in the soil that serve as food for other organisms. Earthworms can weigh between 100 and 1,000 kg ha^{-1}! Oligochaeta, a subclass of annelids (segmented worms), contribute to the texture of soil. The earthworms burrow throughout the soil creating tunnels and shafts. These openings loosen the soil structure providing spaces by which other organisms can travel and gain access to other areas. Water also fills these spaces, leading to water absorption by the soil, which many organisms need, including earthworms. When water gets absorbed into the ground there is less runoff on the surface. While the worms burrow, they take in soil and dead plant material down the profile. Digesting these nutrient-rich materials, they produce waste called *castings*. The castings contain all the nutrients (N, K, Ca, P and Mg) that were present in the original organic material and contain microorganisms, such as bacteria and fungi. The protozoa and nematodes feed on the microorganisms within the castings and release the nutrients within them. The worm castings are valued as manures because they help to balance the pH levels, retain moisture, improve drainage and control pathogens.

As they move through soil, earthworms secrete cutaneous mucus, a source of plant-available nutrients, which is metabolized by soil microorganisms. Earthworm-processed soil contains carbon enriched, earthworm-specific soil aggregates (e.g. casts and middens) in comparison to soil not inhabited by earthworms. The organic polymeric substances in earthworm mucus bind to soil minerals. Mucus from anecic (*Lumbricus terrestris* L.) and endogeic (*Apporectodea caliginosa* Sav.) earthworm species, as characterized by FTIR and ^{13}C-NMR spectroscopy, is dominated by proteins and carbohydrates. Between 21 and 36% of the total organic carbon in the mucus is adsorbed to soil minerals such as illite and goethite and this leads to the formation of mucus-mineral associations. These associations contribute to the retention of organic substances (from earthworm mucus) in soil micro-aggregates and explain the altered physicochemical properties of earthworm-formed aggregates in comparison to the earthworm unaffected "bulk" soil material.

There are also microscopic worms called *nematodes*, or roundworms. These worms live in the water around soil particles. There are several different types of nematodes, some of them eat dead materials, while others eat living roots and some eat other living organisms. Some nematodes can cause severe root damage. Other macro-fauna includes mollusks and arthropods. Arthropods have exoskeletons and jointed legs and these include mites, millipedes, centipedes, springtails and grubs.

Thus, soil macro-fauna includes a wide variety of animals that help maintain a balance in the food web. Some of these consume live or dead plant material and some are predators of the microorganisms mentioned above. In addition, macro-fauna help maintain natural population sizes of their prey.

27.5. Factors Affecting Microbial Population in Soil

- Moisture
- Temperature
- Light
- Air
- Food supply
- Energy
- Nature of soil
- Soil reaction

Microbes can grow and multiply only in the presence of water. Water serves as a source of nutrients to microbes. Excess moisture in soil excludes air in soil and the aerobic organisms are suppressed, while anaerobes take the upper hand.

The suitable range of temperature varies with each organism and accordingly, there are three groups of organisms. Those that thrive below 20 ºC are *psychrophiles*; those that thrive at temperatures between 20 ºC and 40 ºC are *mesophiles*; those that can tolerate high temperatures are called *thermophiles*. Direct sunlight may kill most microbes. Diffused light, however, is useful for algae.

Microbes take their O_2 mainly from soil air. In the absence of soil air, the microbes take O_2 from the inorganic compounds rich in O_2 (such as nitrates, sulphates, etc.). During respiration, microbes take up O_2 and release CO_2. The rate of absorption of O_2 and the rate of release of CO_2 is often taken as a measure of the activities of the microbes.

Almost all microbes derive their food from the soil itself. Most of them, with the exception of algae and some protozoa, are *heterotrophic*. They obtain CO_2 from soil organic matter and other nutrients from the soil mineral matter, water and air.

Most soil microbes, with the exception of photosynthetic algae, obtain their energy from organic materials added to the soil. They utilize the energy released during the oxidation of various compounds that constitute soil organic matter. Thus, organic matter acts both as a source of energy and as a source of nutrients.

The physical and chemical nature of the soil affects the microbes, both quantitatively and qualitatively. For instance, the process of N_2 fixation takes place in the presence of Mo. Lack of Mo in the soil restricts the growth of *Azotobacter*, and, therefore, retards N_2 fixation.

Most microorganisms need an optimum pH range for survival and function. Soil reaction has a definite influence on the growth of microbes in soil. An acid soil usually favours the development of algae (pH 4.5 to 5.5). Most bacteria and protozoa prefer neutral soils (pH 6.5 to 7.5). Actinomycetes require a slightly alkaline reaction.

Table 27.1: The range and optimum pH values for microbes

Microorganisms	Range	Optimum
Bacteria	5 - 9	7
Actinomycetes	6.5 - 9.5	8
Fungi	2 - 7	5
Blue green bacteria	6 - 9	>7
Protozoa	5 - 8	>7

27.6. Role and Functions of Soil Organisms

As is described above, the soil organisms contribute greatly to the wellness of the plants that grow from the soil. In a handful of fertile soil, there are more individual organisms than the total number of human beings that have ever existed. They are dependent on each other to ensure that the environment stays well-balanced. Each type of organism in the soil plays a role in keeping the diverse soil ecosystem worthy of our attention, as depicted in Fig. 27.1.

Soil organisms are not always beneficial. Fungi and bacteria have been known to pose many problems by forming parasitic relationships with plants, attacking young or damaged roots and inhibiting their growth. Macro-fauna is also not faultless. Nematodes are potential parasites, feeding on plants and animals, including humans. They spread diseases and viruses amongst soil organisms, posing many problems to farmers. Larger nematodes are often introduced as predators to smaller nematodes to control their population. Symphylans, the soil-dwelling arthropods, are pests in greenhouses and gardens, feeding on plants and microorganisms. Mites pose similar concerns.

Functions of Soil Organisms

Types of soil Organism		Major Functions
Photosynthe-sizers	• Plants • Algae • Bacteria	**Capture energy** • Use solar energy to fix CO_2. • Add organic matter to soil (biomass such as dead cells, plant litter, and secondary metabolites).
Decomposers	• Bacteria • Fungi	**Break down residue** • Immobilize (retain) nutrients in their biomass. • Create new organic compounds (cell constituents, waste products) that are sources of energy and nutrients for other organisms. • Produce compounds that help bind soil into aggregates. • Bind soil aggregates with fungal hyphae. • Nitrifying and denitrifying bacteria convert forms of nitrogen. • Compete with or inhibit disease-causing organisms.
Mutualists	• Bacteria • Fungi	**Enhance plant growth** • Protect plant roots from disease-causing organisms. • Some bacteria fix N_2. • Some fungi form mycorrhizal associations with roots and deliver nutrients (such as P) and water to the plant.
Pathogens Parasites	• Bacteria • Fungi • Nematodes • Micro arthropods	**Promote disease** • Consume roots and other plant parts, causing disease. • Parasitize nematodes or insects, including disease-causing organisms.
Root-feeders	• Nematodes • Macroarthropods (e.g., cutworm, weevil larvae, & symphylans)	**Consume plant roots** • Potentially cause significant crop yield losses.
Bacteri-al-feeders	• Protozoa • Nematodes	**Graze** • Release plant available nitrogen (NH_4+) and other nutrients when feeding on bacteria. • Control many root-feeding or disease-causing pests. • Stimulate and control the activity of bacterial populations.
Fungal-feeders	• Nematodes • Microarthropods	**Graze** • Release plant available nitrogen (NH_4+) and other nutrients when feeding on fungi. • Control many root-feeding or disease-causing pests. • Stimulate and control the activity of fungal populations.
Shredders	• Earthworms • Macro arthropods	**Break down residue and enhance soil structure** • Shred plant litter as they feed on bacteria and fungi. • Provide habitat for bacteria in their guts and fecal pellets. • Enhance soil structure as they produce fecal pellets and burrow through soil.
Higher-level predators	• Nematode-feeding nematodes • Larger arthropods, mice, voles, shrews, birds, other above ground animals	**Control populations** • Control the populations of lower trophic-level predators. • Larger organisms improve soil structure by burrowing and by passing soil through their guts. • Larger organisms carry smaller organisms long distances.

Fig. 27.1: Functions of soil organisms

28

Functions of Soil

Although plants can be grown to maturity in aerated nutrient solutions and other media, virtually all land plants, including those on which we depend for food, fibre and fuel, grow in soil. The soil has three-fold function to perform with respect to crop growth: physical, chemical and biological.

28.1. Physical Functions of Soil

1. Acts as mechanical support and anchorage for the growing plant.
2. Acts as a reservoir of water and air for plant roots to absorb water and respire.
3. Soil stores sun's heat and supplies to growing seedlings.
4. Soil texture and structure help plants overcome problems of aeration and water content.

28.2. Chemical Functions of Soil

1. Acts as a store-house of plant nutrients. Soil is the source of 13 of the 16 essential plant nutrients.
2. Mineral matter (derived from the weathering of rocks and minerals) and organic matter (from decaying plant and animal residues) provide nutrients for plants.
3. Soil binds and holds nutrients in its fold for subsequent uptake by crop plants.

28.3. Biological Functions of Soil

1. Soil is a habitat for a very large number of plant and animal species.
2. Worms aerate the soil and disintegrate and mix soil constituents by passing large quantities of soil through their bodies and later ejecting them as worm-casts.
3. Burrows made by organisms as insects, rats, moles, etc., facilitate entry of air and water into lower layers of soil, thus helping aeration and drainage.

4. Microbes decompose the dead plant and animal matter and convert complex proteins, celluloses, starch, etc., into simpler substances as nitrates, etc., which serve as nutrients for growing plants.
5. Microbes immobilize simple inorganic nutrients into organic molecules and thus help prevent loss of nutrients from the soil through erosion, etc.

The various physical, chemical and biological changes impart a dynamic character to the soil. This is in sharp contrast to the static nature of rocks and minerals from which it was derived. The magnitude of plant growth reflects a composite of favourable factors such as adequate aeration, water, nutrients, adequate soil depth and proper temperature.

28.4. Ecosystem Functions of Soil

Soils deliver several ecosystem services that facilitate life on earth, as depicted in the Fig. 28.1. An overview of the ecosystem functions of the soil in the environment is presented in Chapter 30.

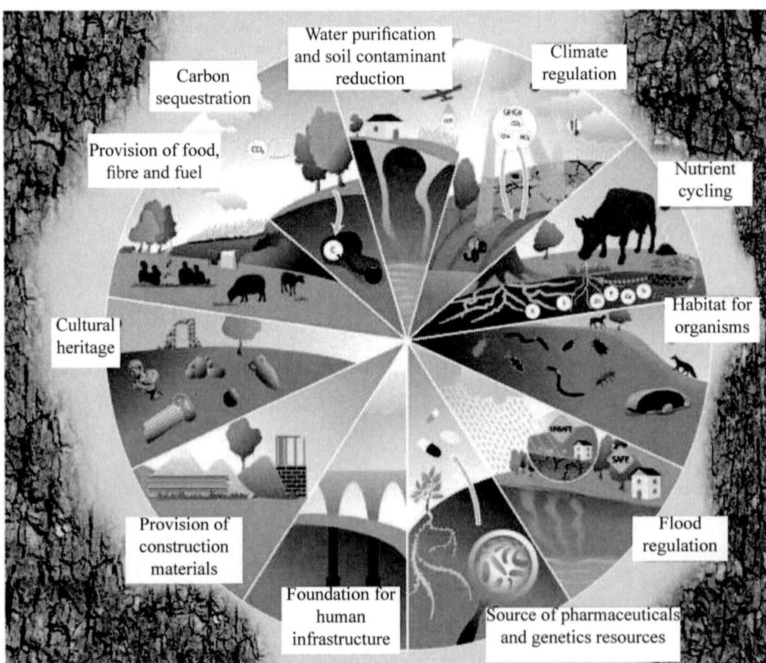

Fig. 28.1: The multifarious ecosystem functions of soil

29

Soil Pollution

Soil pollution indicates the presence in the soil of any chemical or substance out of place and/or present at an abnormally higher concentration so as to adversely affect any non-targeted organism. Soil pollution often cannot be visually perceived or directly assessed, thus rendering it a latent danger. The Status of the World's Soil Resources Report identified soil pollution as one of the main soil threats affecting global soils and the ecosystems services provided by them. Concerns about soil pollution are growing in every region. Recently, the United Nations Environmental Assembly (UNEA-3) adopted a resolution calling for accelerated actions and collaboration to address and manage soil pollution. This consensus, achieved by more than 170 countries, is a clear sign of the global relevance of soil pollution and of the willingness of these countries to develop concrete solutions to address the causes and impacts of this major threat.

29.1. Sources of Soil Pollution

The main anthropogenic sources of soil pollution are the chemicals used in or produced as byproducts of industrial activities, domestic, livestock and municipal wastes (including wastewater), agrochemicals, and petroleum-derived products. These chemicals are released to the environment accidentally, for example from oil spills or leaching from landfills, or intentionally, as is the case with the use of fertilizers and pesticides, irrigation with untreated wastewater, or land application of sewage sludge. Soil pollution also results from atmospheric deposition from smelting, transportation, pesticide applications and combustion of many substances as well as radionuclide fallouts from atmospheric weapons testing and nuclear accidents. New concerns are being raised about emerging pollutants such as pharmaceuticals, endocrine disruptors, hormones and toxins, among others, and biological pollutants, such as micropollutants in soils, which include bacteria and viruses.

29.2. Effects of Soil Pollution

Soil pollution can severely degrade the major ecosystem services provided by soil. It decreases food security by reducing crop yields due to toxic levels of contaminants and by rendering produce from polluted soils unsafe for consumption by humans and animals. Many contaminants (including major nutrients such as nitrogen and phosphorus) are transported from the soil to surface waters and ground water, causing great environmental harm through eutrophication and direct human health issues. Pollutants also directly harm soil microorganisms and other soil fauna, thus affecting soil biodiversity and the services provided by the affected organisms. Risks to human health from soil pollution arise from heavy metal contamination such as arsenic, lead, and cadmium, as well as from organic chemicals such as polychlorinated biphenyls, polycyclic aromatic hydrocarbons and pharmaceuticals. The health risks associated with the widespread soil contamination by radionuclides from the Chernobyl disaster in 1986 are an enduring memory for many people.

Soil pollution causes a chain reaction. It alters biodiversity and reduces the capacity of soil to act as a filter. It also contaminates the water stored in the soil and groundwater, and causes an imbalance of soil nutrients. Among the most common soil pollutants are heavy metals, persistent organic pollutants and emerging pollutants like pharmaceutical and personal care products.

Soil pollution induces devastating consequences to the environment and to all forms of life. Unsustainable agricultural practices that reduce soil organic matter facilitate the transfer of pollutants into the food chain. Thus, in polluted soil, the contaminants accumulate in plant produce which are passed on *via* grazing animals and birds to the humans who consume the plants and animals. Pollutants in soil, groundwater and in the food chain can cause several diseases and excess mortality in humans, from short-term acute effects such as intoxications or to long-term chronic effects like cancer.

Soil pollution also has high economic costs due to the reduction of crop yields and crop quality. The fact that the vast majority of pollutants result from human action means that we are directly responsible for a secure future. Soils need to be recognized for their productive capacities as well as for their contribution to food security and maintenance of key ecosystem services.

29.3. Fallouts of Soil Pollution

The following are just a few reasons why soil pollution cannot be ignored:

1. **Soil pollution affects everything.** The food we eat, the water we drink, the air we breathe, our health and the health of all the organisms on Earth is dependent on healthy soil. The nutrient content of a plant's tissues is directly related to the nutrient content of the soil and its ability to exchange nutrients and water with the plant roots.

2. **Soil pollution is invisible.** Today, a third of our soils are moderately to severely degraded due to erosion, loss of soil organic carbon, salinization, compaction, acidification and chemical pollution. Nature takes about 1,000 years to form 1 cm of top soil, meaning that we won't be able to produce more soil within our lifetime. The current rate of soil degradation endangers the capacity of future generations to meet their most basic needs. It must be borne in mind that soil is a precious, non-renewable natural resource.
3. **Soil pollution affects soils' capacity to filter.** Soils act as a filter and buffer for contaminants. The potential of soil to cope with the strain of pollutants is finite and if this is exceeded, contaminants will trickle into other parts of the environment like our food chain.

29.4. Remediation of Soil Pollution

Remediation of polluted soils is imperative and concerted research continues to develop practicable and sustainable remediation methods. Risk assessment approaches are similar worldwide and consist of a series of steps to be taken to identify and evaluate substances responsible for soil pollution and the severity of risk to the environment and human health. Expensive physical remediation methods such as chemical inactivation or sequestration in landfills are increasingly being replaced by biological methods such as enhanced microbial degradation or phytoremediation. Revised World Soil Charter of FAO advocates that national governments implement strict regulations on soil pollution that preclude accumulation of contaminants beyond established levels to guarantee human health and wellbeing. Governments are also urged to facilitate remediation of contaminated soils that exceed levels established to protect the health of humans and the environment. It is also imperative to limit pollution from agricultural sources by the global implementation of sustainable soil management practices.

30

Soils and The Environment

Soils play a vital role in the health of ecosystems in the world. Soils connect with air, water, rocks and organisms and are responsible for a multitude of functions in the global ecosystem services. These functions include providing habitat for most living things, air composition and quality, temperature regulation, carbon and nutrient cycling, water cycling and quality, and natural waste recycling. Mankind cannot survive without these soil functions.

30.1. Habitat

Soils are the environment in which plants grow. They provide heat, nutrients and water to nurture plants to maturity. These plants, together with other plants and organisms, create ecosystems. These plants then provide valuable habitat and food sources for animals, bacteria and other things. Ecosystems depend on the soil and soils determine the nature and kind of ecosystems.

30.2. Air Quality and Composition

A well covered soil prevents erosion. During the times when wind blows across the soils, fine soil particles are suspended in the air. These are easily inhaled and accumulated in lung tissues causing major respiratory problems. These particles can contain fungi and bacteria, which can further cause infection and diseases.

30.3. Temperature Regulation

Soil temperature plays a key role in many biological and chemical processes taking place in the soil. These include seed germination, activities of microbes that live in the soil environment and decomposition of plant and animal tissues. In the soils of colder regions, less of biological and chemical reactions take place as compared to soils in warmer regions; therefore, there may be more carbon stored in such soils.

30.4. Carbon and Nutrient Cycling

Soil contains nearly 5 times more of stored carbon than the plants that grow in it. Natural processes are all cyclical. The global carbon cycle is more complex and involves carbon stored in fossil fuels, soils, oceans, and rocks. Physical, biological, and chemical processes in the soil regulate the balance in organic carbon compounds, whether they are released to the atmosphere as CO_2, or stored in the soil. The same is the process in case of nitrogen, phosphorus, and all other nutrients.

30.5. Water Filtration

During rains, the soil acts as a sponge, soaking water into the ground. This water can be taken up by plants, microbes and other living beings, or the water moves into the underground aquifers and lakes, and flows into streams before eventually reaching the ocean. If rainfall contains harmful pollutants, the soil acts as a filter, with the soil particles capturing contaminants and the water comes out cleaner in the aquifers and rivers. Such filtering of pollutants by soil takes place through physical, chemical and biological process.

30.6. Water Cycling and Quality

Application of excess fertilizers, particularly nitrogen and phosphorus, can result in runoff to streams and rivers and contamination of groundwater. In most freshwater systems, phosphorus is the major limiting nutrient for the growth of photosynthetic organisms (plants, algae, and some microorganisms like phytoplankton). If a lake or pond receives excess phosphate, it can stimulate these organisms, especially in presence of nitrates and they block out the light and consume oxygen in the water much to the detriment of other organisms as fish. Other symptoms include turbid water that is typically green or yellow which render water bodies unenticing for recreation.

30.7. Waste Decomposition

Soils converts 'waste' products into newer, better things that can be reused by other creatures. When an organism dies, the soil converts the dead material into new materials and food for living things through biological and chemical processes. This is nature's way of recycling.

30.8. Heavy Metals and Radionuclides in Soil

Heavy metals are natural components of the environment warranting concern because they are being added to soil, water and air in increasing amounts. Some micronutrients (as copper, manganese and zinc) are micronutrients essential in small amounts for plant and animal life. They can, however, be harmful if taken up by plants or animals in large amounts, as can other heavy metals not listed as essential nutrients.

Razdionuclides are dangerous due to the radiations they emit. Some radionuclides occur naturally, exposing all living organisms to small and usually harmless doses of radiation, but a majority of them are man-made that accumulate as they pass through a biological cycle. For example, the chemistry of strontium is similar to that of calcium, and radionuclides of strontium accumulate with calcium in bone tissue, where they emit radiation.

30.9. Soils and Global Warming

The mean temperature of the Earth's surface is 288 K (14.8 °C or 58.7 °F). It is kept at this temperature by the greenhouse effect without which it would be about 34 degrees lower. There is concern, though, that higher concentrations of CO_2 and other gases in the atmosphere will lead to global warming. There is also concern that the emission of certain gases will decrease the layer of ozone in the stratosphere, which will lead to more ultraviolet light reaching the Earth's surface.

Distinction has to be made between a greenhouse effect which keeps the Earth's surface warm at a roughly constant temperature and an enhanced greenhouse effect which results in global warming. The greenhouse effect is benign, but global warming is expected to lead to rising sea levels, changes in annual and seasonal rainfall at a regional scale and to effects on vegetation systems and agriculture. Rising sea levels will require substantial capital investment to prevent flooding and might cause serious loss of life. Changes in weather patterns might lead to food shortages. Although emission of CO_2 from the burning of fossil fuels is the main cause of the problem, there is also an appreciable flux of CO_2 from the oxidation of soil organic matter and the burning of forests.

30.10. Soils in the Environment: Problems and Solutions

Soils are needed to grow crops for food, animal fodder and fibre, trees for fuel and timber, and to support natural woodlands and grasslands. As the human population has increased new land has been brought into cultivation. This trend has ended in many parts of the world where all that remains is land that is either too difficult or uneconomic to cultivate. Conservation of natural ecosystems, the importance of which is being increasingly accepted, further restricts the expansion of agriculture.

Prediction of the growth of the human population is that the population will have reached 10 billion sometime in the second half of this century. Much of this increase will be in the Third World. Sustainable management of soils is the only key to provide this increased population with food and fibre. One pertinent question here will be whether more intensive systems of management can be employed which, at the same time, are sustainable in environments where soils are fragile, the rains are unpredictable and the essential requirements for development are sparse.

There are different problems in countries elsewhere, especially in Europe and North America. Overproduction of food provides some justification for less intensive farming systems, with land being taken out of cultivation, for example, by the set-aside scheme in the UK and the soil bank in the USA; this has led to criticism of the use of fertilizers and pesticides and the advocacy of less intensive systems such as organic farming.

31

Soil Quality and Soil Health

Soil quality is the capacity of the soil to function, within its natural or managed ecosystems, to sustain plant and animal productivity, maintain or enhance water and air quality, support human and animal health and habitation. This is not limited to agriculture, though most work and evaluation has occurred on agricultural lands.

Soil health refers to the condition of the soil and its potential to sustain biological functions, maintain environmental quality and promote plant and animal health. Healthy soils are the key to food security and sustainable future. They support sustainable food production, mitigate and adapt to climate changes, filter water, enhance resilience to floods and droughts and so much more.

31.1. Benefits of a Healthy Soil

The following are benefits of a healthy soil:

- Soils are the basis for healthy food production.
- They host a quarter of our planet's biodiversity.
- Soils help to combat and adapt to climate change.
- They are essential for food security.
- They store and filter water.

31.2. Evaluation for Soil Quality and Health

Soil quality and soil health are of paramount importance for sustainable agro-ecosystem management. Soil quality evaluation is a means to assess management-related changes in the soil and to link existing resource concerns to environmentally sound land management practices. When a soil is to be evaluated for its quality, a set of indicators has to be used. These indicators may be qualitative or quantitative, but should integrate physical and chemical properties and be accessible to many users. They must respond to different management operations and adapt to different climates.

Soil health is an assessment of how well a soil performs all of its functions now and how those functions are being nurtured for future use. Soil health cannot be determined by measuring only crop yield, or water quality, or any other single outcome. For evaluating soil health, a biological indicator is needed. Soil organisms are actively involved in the decomposition of organic matter and cycling of nutrients. An easily discernible biological indicator is the earthworm activity in the soil. In a healthy soil, earthworms are plenty. Earthworms modify the physical structure of soils by producing new aggregates and pores, which improves soil tilth, aeration, infiltration and drainage. They improve soil porosity by burrowing and mixing soil and roots often follow earthworm burrows and use nutrients available with the worm casts. Only 'living' things can have health; so, viewing soil as a living ecosystem reflects the way we care for our soils.

31.3. Soil Quality Indicators

Scientists use soil quality indicators to evaluate how well a soil functions since soil function often cannot be directly measured. It is an effort to identify soil properties that are responsive to management, affect or correlate with environmental consequences and are capable of being precisely measured within certain technical and economic constraints. Soil quality indicators may be qualitative (e.g., drainage is fast) or quantitative (e.g., infiltration = 2.5 in hr^{-1}).

Ideal soil quality indicators should

- correlate well with ecosystem processes.
- integrate soil physical, chemical and biological properties and processes.
- be accessible to many users.
- be sensitive to management and climate.
- be components of existing databases.
- be easily interpretable.

There are three main categories of soil indicators, namely, physical, chemical and biological. Typical soil tests only look at chemical indicators. Soil quality assessment attempts to integrate all three types of indicators. Since the categories do not neatly dovetail with the various soil functions, integration is necessary. The following table displays the relationship between indicator type and soil function.

Indicator category	Related soil function
Physical	Physical stability and support, Water relations, Habitat
Chemical	Nutrient cycling, Water relations, Buffering
Biological	Biodiversity, Nutrient cycling, Filtering

Organic matter, more specifically soil carbon, transcends all three indicator categories and has the most widely recognized influence on soil quality. Organic matter is bound to all soil functions. It influences other indicators, such as aggregate stability (physical), nutrient retention and availability (chemical) and nutrient cycling (biological); and is itself an indicator of soil quality. Some indicators are descriptive and usable in the field as part of a health card. Others must be measured using laboratory analyses. Some examples of indicators that fall into the three broad categories of physical, chemical and biological are furnished below.

31.4. Indicator Categories

Physical indicators furnish information about soil hydrologic characteristics (like infiltration and retention) that influence availability to plants. Some indicators are responsible for nutrient availability by their influence on rooting volume and aeration status. Others tell us about erosional status. Indicators include measures of:

- Aggregate stability
- Available water capacity
- Bulk density
- Infiltration
- Slaking
- Soil crusts
- Soil structure and macropores
- Others...

Chemical indicators offer information about the equilibrium between soil solution (soil water and nutrients) and exchange sites (clay colloids, organic colloids); plant health; nutritional requirements of soil flora and fauna; and levels of soil contaminants and their availability for uptake by flora and fauna. The indicators include measures of:

- Electrical conductivity
- Soil nitrate

- Soil reaction (pH)
- more...

Biological indicators inform us about the soil organisms that form the soil food web that is related to the decomposition of organic matter and nutrient cycling. The numbers of organisms (both individuals and species) that perform similar niches, can indicate functional or resilience capabilities of a soil after a disturbance. Indicators include measures of:

- Earthworms
- Particulate organic matter
- Potentially mineralizable nitrogen
- Respiration
- Soil enzymes
- Total organic carbon
- Others...

31.5. Minimum Data Sets and Indicators

Since it is practically not possible to measure every soil property or ecosystem, researchers have proposed a minimum data set, which is the smallest set of soil properties or indicators needed to measure soil quality. A minimum data set is established by identifying key soil properties or attributes that are sensitive to change in soil functions. Table 31.1 gives an example of a minimum data set, showing the relationship of each indicator to soil health concerns.

Table 31.1: A typical minimum data set of indicators for soil quality

Indicators	Relationship to soil health
1. Physical Indicators	
Soil structure	Retention and transport of water and nutrients, habitat for microbes and soil erosion
Depth of soil and rooting	Estimate of crop productivity potential, compaction and plough pan
Bulk density and infiltration	Water movement, porosity and workability
Water holding capacity	Soil water storage and availability
2. Chemical Indicators	
pH	Biological activity and nutrient availability
Electrical conductivity	Plant growth, microbial activity and salt tolerance
Soil organic matter	Soil fertility, structure, aggregate stability, nutrient retention, soil erosion and available water capacity
Soil available N, P and K	Plant available nutrients
3. Biological Indicators	
Microbial biomass C and N	Microbial catalytic potential and repository for C and N
Potentially mineralizable N	Soil productivity and N supplying potential
Soil respiration	Microbial activity measure

A minimum data set does not usually include all relevant properties for a region or farming system. Nevertheless, it is an example of a minimum set of indicators required to obtain a comprehensive understanding of the soil evaluated. Each minimum data set is tailored to a particular region or soil map unit (soil type) and includes only those properties relevant to the soil types, farming system and land uses of the areas being evaluated. For example, a minimum data set for the inland regions would probably not include such indicators as salt accumulation and electrical conductivity, while a data set for areas with arid and semi-arid soils would include these indicators. Compiling a minimum data set helps to identify locally relevant indicators and to evaluate the link between indicators selected and the significant soil and plant properties for the region.

32

Soils of India

India is endowed with varied types of soils and this could be attributed varied relief features, landforms, climatic realms and vegetation types in the subcontinent. In ancient times, soils used to be classified in India into fertile and sterile soils. Later, the inherent characteristics and external features of soils such as texture and colour, were used to classify the soils. Based on texture, soil types were identified as sandy, clayey, loam, etc. In terms of colour, they were red, yellow, black, etc.

Scientific surveys of soils in India have been conducted by various agencies since independence. All India Soil Survey Organization, established in 1956, carried out detailed soil surveys in the catchment areas of major river valley projects like the Damodar Valley. Later, the National Bureau of Soil Survey and the Land Use Planning (NBSSLUP) of the Indian Council of Agricultural Research, established in 1976, made extensive studies on Indian soils and classified them as per the USDA Soil Taxonomy (Table 32.1).

Table 32.1: Soils of India

S. No.	Order	Area (in 10^3 ha)	Percentage
1.	Inceptisols	130372.90	39.74
2.	Entisols	92131.71	28.08
3.	Alfisols	44448.68	13.55
4.	Vertisols	27960.00	8.52
5.	Aridisols	14069.00	4.28
6.	Ultisols	8250.00	2.51
7.	Mollisols	1320.00	0.40
8.	Others	9503.10	2.92
	Total		100.00

Source: Soils of India, National Bureau of Soil Survey and Land Use Planning, Publication Number 94

On the basis of genesis, colour, composition and location, the soils of India have generally been classified thus:

i. Alluvial soils
ii. Black soils
iii. Red and Yellow soils
iv. Laterite soils
v. Arid soils
vi. Saline soil
vii. Peaty soils
viii. Forest soils

32.1. Alluvial Soils

Alluvial soils are found in the northern plains and the river valleys, covering about 40 per cent of the total area of the country. These soils are transported and deposited by rivers and streams. In the Peninsular India, they are spread in deltas of the east coast and in the river valleys.

The alluvial soils vary in texture from sandy loam to clay, light grey to ash grey in colour and generally rich in potash but poor in phosphorous. These soils are loamy and clayey in the lower and middle Ganga plain and the Brahmaputra valley. Alluvial soils are intensively cultivated.

32.2. Black Soils

Black soils cover most of the Deccan Plateau which includes parts of Maharashtra, Madhya Pradesh, Gujarat, Andhra Pradesh and some parts of Tamil Nadu. The black soils are very deep in the upper Godavari and the Krishna and in the north western part of the Deccan Plateau. These soils are traditionally known as the '*Regur* Soils' or 'Black Cotton Soils'.

The black soils are generally deep, clayey, deep black to grey in colour and practically impermeable. They swell and are sticky during monsoons. During summer, these soils develop wide cracks, leading to *'self-ploughing'*. Because of poor percolation, the black soils retain moisture for a very long time, which helps especially the rain-fed crops to survive even during the dry season. Generally, the black soils are rich in lime, iron, magnesium and potash, but poor in phosphorous, nitrogen and organic matter.

32.3. Red and Yellow Soils

Red soils are found in areas of low rainfall in the eastern and southern part of the Deccan Plateau. Long stretch of areas along the piedmont zone of the Western Ghats are red loamy soils. Yellow and red soils are also prevalent in the southern parts of the middle Ganga plain and in parts of Odisha and Chattisgarh. The soil is reddish in colour due to iron in crystalline igneous and metamorphic rocks which is yellow when in hydrated form. The fine-grained red and yellow soils are generally fertile, but the coarse-grained soils found in dry upland areas are poor in fertility. These soils are poor in nitrogen, phosphorous and organic matter.

32.4. Laterite Soils

Laterite soils owe their name to the Latin word *'Later'* which means 'brick'. These soils develop in areas with high temperature and high rainfall and as a result of intense leaching due to tropical rains. Lime and silica are leached away, rendering the profile rich in and iron and aluminium oxides. Humus is decomposed fast by the microbes that thrive well in tropical conditions. The laterite soils are poor in organic matter, nitrogen, phosphate and calcium, but rich in iron and potash. Red laterite soils of Tamil Nadu, Andhra Pradesh and Kerala are more suited for tree crops like cashew. The laterite soils are common in Karnataka, Kerala, Tamil Nadu, Madhya Pradesh and the hilly areas of Odisha and Assam.

32.5. Arid Soils

Arid soils are characteristically developed in western Rajasthan which exhibit typically arid topography. The colour of arid soils range from red to brown. They are generally sandy and saline in nature. Because of the extreme dry climate, high temperature and high evaporation, these soils lack in moisture, nitrogen and humus. The phosphate content is normal. The calcium content increases down the profile and *kankars* are found in the lower horizons. The *kankar* layers in the lower horizons restricts the percolation of water and as such when irrigation is available, the water is readily available for crop growth.

32.6. Saline Soils

Also known as *usara* soils, saline soils contain a larger proportion of carbonates, bicarbonates, chlorides and sulphates of sodium, calcium, potassium and magnesium. They have more salts, largely because of the dry, arid and semi-arid climate and poor leaching. They are not fertile as they lack in all major plant nutrients and do not support any vegetative growth. Saline soils are widespread in western Gujarat, deltas of the eastern coast and in the Sundarbans of West Bengal. Seawater intrusions lead to the formation of saline soils. With excessive use of irrigation, the fertile alluvial soils in of intensive cultivation areas are turning saline. Excessive irrigation under dry climatic conditions induces capillary action, resulting in the deposition of salts on top soil.

32.7. Peaty Soils

In areas of heavy rainfall and high humidity, where there is a good growth of vegetation, peaty soils are found. Large chunk of dead organic matter accumulates in these areas, giving a rich humus and organic content and black colour to the soil. Organic matter content in these soils may be even 40-50 per cent. Such soils occur widely in the northern parts of Bihar, southern parts of Uttaranchal and the coastal areas of West Bengal, Orissa and Tamil Nadu.

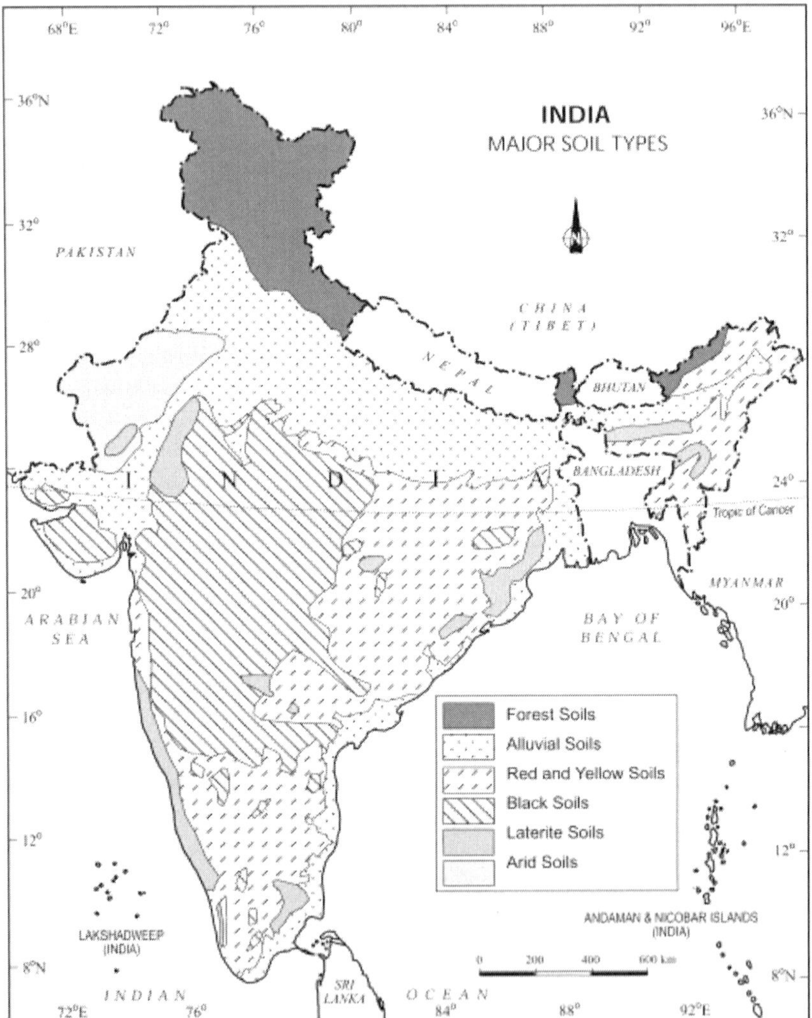

Fig. 32.2: Major soil types of India

32.8. Forest Soils

Forest soils are formed typically in the forest areas served by adequate rainfall. Their texture and structure are defined by the mountainous environment in which they are formed. They are coarse-grained in the upper slopes and loamy and silty on valley sides and. In the snow-clad areas of the Himalayas, they experience denudation, and are acidic with low humus content. The soils in the lower valleys are fertile.

Multiple Choice Questions

1. The father of Pedology is:
   (a) Sorenson (b) Hilgard
   (c) Dokuchaev (d) Kononova
2. The science dealing with the laws of origin, formation and geographic distribution of soil as a body in nature is:
   (a) Pedology (b) Edaphology
   (c) Geology (d) Paleontology
3. The study of soils from the standpoint of higher plants is called:
   (a) Edaphology (b) Soil Taxonomy
   (c) Geology (d) Pedology
4. The most dominant element on earth's crust is:
   (a) Carbon (b) Oxygen
   (c) Silicon (d) Hydrogen
5. The most dominant metal in earth's crust is silicon:
   (a) Carbon (b) Iron
   (c) Aluminium (d) Silicon
6. Pick out the *incorrect* statement about soil and soil formation:
   (a) Pedon is the smallest volume of soil recognizable in nature.
   (b) Soil is a two-dimensional body.
   (c) Weathering is a process of disintegration of rocks and minerals.
   (d) Pedogenic processes are dynamic and extremely complex in nature.

7. Pick out the *correct* statement about rocks and soil formation:
   (a) High temperature and pressure can change a pre-existing igneous rock into a new metamorphic rock.
   (b) In arctic regions chemical weathering is intense and in humid tropics physical weathering is intense.
   (c) The rocks that are formed when the molten magma is cooled and consolidated beneath the earth's surface are called extrusive rocks.
   (d) None of the above.

8. The rocks that are formed when the molten magma is cooled and consolidated beneath the earth's surface are called:
   (a) Plutonic rocks (b) Volcanic rocks
   (c) Sedimentary rocks (d) Metamorphic rocks

9. Rocks formed on the surface of the earth are termed as:
   (a) Intrusive (b) Volcanic
   (c) Sills (d) Dykes

10. The rocks that cover 75% of earth's surface are:
    (a) Igneous rocks (b) Sedimentary rocks
    (c) Metamorphic rocks (d) None of the above

11. The example for a sedimentary rock is:
    (a) Sandstone (b) Limestone
    (c) Shale (d) All the above

12. Which of the following statements about sedimentary rocks is *false*?
    (a) Sedimentary rocks cover 75% of the earth's surface.
    (b) Sedimentary rocks make up less than 5% of all the rocks that make up Earth.
    (c) Sedimentary rocks are formed in layers.
    (d) Sedimentary rocks are always formed at a place far away from the parent igneous rocks from which they are formed.

13. Jagged rock bits which are cemented are called:
    (a) Conglomerate (b) Shale
    (c) Sandstone (d) Porphory

14. Sedimentary particles with clay size particles are called as:

(a) Argillaceous (b) Arenaceous

(c) Rudaceous (d) None of the above

15. The sedimentary rock that undergoes metamorphism and becomes slate is:

(a) Sandstone (b) Limestone

(c) Shale (d) Porphory

16. The metamorphic rock marble is formed from:

(a) Granite (b) Sandstone

(c) Limestone (d) Shale

17. A natural substance having a definite chemical composition and commonly a definite molecular arrangement is called:

(a) Rock (b) Mineral

(c) Soil (d) All the above

18. The most important group of essential minerals in soil which are alumino-silicates:

(a) Pyroxenes (b) Zeolite

(c) Feldspars (d) Quartzite

19. An accessory mineral that is not an oxide, hydroxide, sulphide or carbonate is:

(a) Corundum (b) Galena

(c) Malachite (d) Apatite

20. The minerals comprising the bulk of the rocks are called:

(a) Primary minerals (b) Accessory minerals

(c) Clay minerals (d) Amorphous minerals

21. The soils materials transported by the force of gravity form:

(a) Colluvium (b) Alluvium

(c) Loess (d) All the above

22. Iron containing mineral is:

(a) Bauxite (b) Hematite

(c) Apatite (d) Calcite

23. All the minerals in the following list belong to a particular group, except one. Which one is that?

(a) Olivine (b) Feldspar

(c) Muscovite (d) Clay

24. The process of disintegration and decomposition of rocks and minerals is:

(a) Physical weathering (b) Chemical weathering

(c) Biological weathering (d) All the above

25. Physical weathering is a kind of mechanical disruption and is termed as:

(a) Disintegration (b) Decomposition

(c) Double decomposition (d) Hydrolysis

26. Biological weathering is responsible for:

(a) Decomposition of rocks and minerals

(b) Disintegration of rocks and minerals

(c) Both decomposition and disintegration

(d) Neither decomposition nor disintegration

27. The "weathering" of rocks and minerals refers to:

(a) Their physical breakup into smaller particles and their chemical alteration into dissolved ions and new types of minerals.

(b) The formation of soil horizon distinctions because of the effects of rain, snow, freezing temperatures, heat, atmospheric pressure, and other weather-related conditions.

(c) The joining of basic elements in the earth's crust to form the minerals found in igneous rocks.

(d) All the above are correct.

28. The most resistant mineral in the earth's crust is:

(a) Calcite (b) Quartz

(c) Olivine (d) Pyroxene

29. As the particle size decreases, chemical weathering:

(a) Decreases (b) Increases

(c) First decreases and then increases (d) Remains constant

30. The mixing of soil by animals such as ants, earthworms, moles, rodents, and man himself is called:

(a) Floral pedoturbation (b) Faunal pedoturbation

(c) Argillic pedoturbation (d) None of the above

31. The phenomenon of the surface layer peeling off from the parent mass, resulting ultimately in the disintegration of a rock is called:

(a) Abrasion (b) Disintegration

(c) Fragmentation (d) Exfoliation

32. The process of formation of soils from weathered parent materials is called:

(a) Physical weathering (b) Chemical weathering

(c) Pedogenesis (d) Geochemical weathering

33. The scientist who established that soils develop as a result of the action of soil forming factors climate, organisms, parent material, time and relief is:

(a) Joffe (b) Jenny

(c) Dokuchaev (d) Hilgard

34. A passive soil forming factor is:

(a) Climate (b) Organism

(c) Parent material (d) All the above

35. Which of the following are **active** factors of soil formation?

(a) Time (b) Topography

(c) Parent material (d) None of the above

36. The three passive factors of soil formation are:

(a) Climate, Organisms, Time (b) Climate, Relief, Time

(c) Parent material, Relief, Time (d) Parent m aterial, Organisms, increases

37. Mechanical migration of clay and silt particles from A to B horizon is:

(a) Lessivage (b) Infiltration

(c) Percolation (d) Eluviation

38. The synthesis of humus from leaf litter is an example of which basic soil-forming process?

(a) Additions (b) Losses

(c) Translocations (d) Transformations

39. The process that operates most favorable in warm and humid (tropical) climate with high rainfall is:

(a) Podzolization (b) Laterization

(c) Gleization (d) Solodization

40. If the parent material and period of soil formation are similar, we would expect to find the most highly weathered soils in:

(a) Cool, dry climate regions (b) Hot, humid tropical regions

(c) Cool, wet, valley swamps (d) Hot, windy, desert regions

41. Biological, physical churning of soil materials, homogenizing solum in varying degrees is called:

(a) Pedoturbation (b) Illuviation

(c) Eluviation (d) Chelation

42. The process that operates most favorable in cool, temperate climate with coniferous vegetation is:

(a) Podzolization (b) Laterization

(c) Gleization (d) Solodization

43. Laterization is a process that removes from the upper layers:

(a) Sesquioxides alone (b) Silica alone

(c) Both silica and sesquioxides (d) Neither silica nor sesquioxides

44. The exposed vertical section of a soil showing different layers from the surface is:

(a) Soil profile (b) Control section

(c) Solum (d) Polypedon

45. The underlying consolidated bedrock in a soil profile is:

(a) Intrusive rock (b) Sedimentary rock

(c) Parent material (d) Regolith

46. A soil profile has the following horizons: Oi, Oa, A1, A2, AB, B, B1, B2, BC, CB, C. How many *master* horizons does it have?

(a) Eleven (b) Nine

(c) Four (d) Three

(e) Two

47. In a soil that contains all of these horizons, which would be at the bottom?

(a) A horizon (b) B horizon

(c) C horizon (d) O horizon

48. Which of the following horizons is usually considered to be "subsoil"?

(a) A horizon (b) B horizon

(c) C horizon (d) O horizon

49. The regolith would likely include which horizon(s)?

(a) A horizon (b) E horizon

(c) B horizon (d) C horizon

50. Compared to the B horizon, the A horizon is likely to be higher in:

(a) Clay (b) Organic matter

(c) Sesquioxides (d) All the above

51. The notation 'n' is used as suffix to designate specific subordinate distinction within the master horizon to indicate the accumulation of:

(a) Nitrogen (b) Carbonates

(c) Silica (d) Sodium

52. Which of the following is *not* correctly matched?

(a) O Horizon – Organic Horizon

(b) A_2 Horizon – Horizon of eluviation

(c) B_1 Horizon - Transition between B and C

(d) C Horizon – Horizon below the solum

53. Regolith refers to:

(a) Solum
(b) Pedon
(c) Consolidated bed rock
(d) Loose, unconsolidated solid material covering the bed rock

54. The notation B_t refers to the accumulation (in the B horizon) of:

(a) Sand
(b) Silt
(c) Clay
(d) Ca carbonate

55. Which of the following is regarded as a soil individual?

(a) Pedon
(b) Polypedon
(c) Profile
(d) Horizon

56. Identify the **master** horizon from among the following horizons:

(a) A
(b) a
(c) AE
(d) None of the above

57. Identify the **subordinate** horizon from among the following horizons:

(a) A
(b) t
(c) AE
(d) None of the above

58. Identify the **transition** horizon from among the following horizons:

(a) O
(b) t
(c) AE
(d) None of the above

59. A hardpan horizon:

(a) Petrocalcic horizon
(b) Petrogypsic horizon
(c) Plinthite
(d) All the above

60. Soil taxonomy is based on which approach?

(a) Qualitative
(b) Comparative
(c) Quantitative
(d) Relative

61. Soils developed from volcanic ash belong to the order:

(a) Andisol
(b) Gelisol
(c) Vertisol
(d) Aridisol

62. Soils with poorly developed profile because of time as limiting factor are called as:

(a) Zonal soils (b) Azonal soils

(c) Intrazonal soils (d) None of the above

63. Periyanaickenpalayam clayey, calcareous, isohyperthermic, mixed Typic Ustivertept. In this taxonomic description, the word 'Typic' denotes:

(a) Order (b) Suborder

(c) Great group (d) Subgroup

64. A mineral horizon that has >50% base saturation, >1% organic matter and developed in grasslands is:

(a) Umbric epipedon (b) Mollic epipedon

(c) Histic epipedon (d) Ochric epipedon

65. The soil temperature regime is **thermic** when the mean annual soil temperature is:

(a) >8 - <15 °C (b) >15 °C, but <22 °C

(c) >22 °C (d) None of the above

66. To be designated as clay a soil must contain at least this amount of clay separates:

(a) 15% (b) 35%

(c) 55% (d) 75%

67. The type of texture that is easy to handle in tillage operations is:

(a) Sandy (b) Silty

(c) Clayey (d) None of the above

68. An ideal soil will have a pore space of:

(a) 25 % (b) 50 %

(c) 75 % (d) 100 %

69. The clay particles have diameter less than:

(a) 0.002 mm (b) 20 micro metre

(c) 0.00002 cm (d) 0.00002 m

70. The 'feel' of the soil is represented by:

(a) Texture (b) Structure

(c) Consistence (d) All the above

71. If the mineral fraction of a soil is composed of 40% sand, 30% silt and 30% clay, what is its textural class?

(a) Clay loam (b) Silty clay loam

(c) Silt loam (d) Sandy clay loam

72. If the mineral fraction of a soil is composed of 50% sand, 10% silt and 40% clay, what is its textural class?

(a) Sandy clay loam (b) Sandy clay

(c) Sandy loam (d) Clay loam

73. Three particles have a diameter of 0.30, 1.80 and 0.06 mm. The three particles are:

(a) Sand, sand and sand (b) Silt, sand and silt

(c) Clay, silt and clay (d) Silt, sand and clay

74. The 'architecture' of the soil is represented by:

(a) Texture (b) Structure

(c) Consistence (d) All the above

75. Flocculation is brought about by coalescence of colloidal particles, facilitated by:

(a) Phosphorus (b) Potassium

(c) Calcium (d) Sodium

76. If the corners and edges of peds are rounded, the structure is described as:

(a) Angular (b) Subangular

(c) Granular (d) Structureless

77. The ion responsible for deflocculation in soils is:

(a) Na^+ (b) Ca^{2+}

(c) Mg^{2+} (d) All the above

78. To avoid damaging the soil structure, the best time to perform tillage on a soil is when it is:

(a) Very dry and hard (b) Very wet and soft

(c) Moist and friable (d) Hot and dry

79. Soil strength is likely to be relatively high for soils that are:

(a) Hard and dry (b) High in clay

(c) High in bulk density (d) All the above

80. The capacity of soil to being molded when the soil mass is subjected to a deforming stress is called:

(a) Adhesion (b) Cohesion

(c) Plasticity (d) Adsorption

81. Which of the following is not a phenomenon of soil consistence?

(a) Plasticity (b) Friability

(c) Stickiness (d) Aggregation

82. Which of the following statements is wrong?

(a) Soils rich in kaolinite clay exhibit better consistence than soils rich in montmorillonite.

(b) Cohesiveness increases with decreasing particle size.

(c) A puddled soil has more cohesion than a well-aggregated soil.

(d) All the above statements are wrong.

83. Which of the following terms is not used to describe soil consistence?

(a) Friable (b) Plastic

(c) Sticky (d) Pliable

84. If the soil sticks together only lightly and is soft, the consistence is called:

(a) Friable (b) Plastic

(c) Sticky (d) None of the above

85. A large pot is filled with 12 litres of a mineral soil. Its bulk density was determined to be 1.25 g cm^{-3}. What is the volume of total pore space contained in the 12 litres of soil? [Assume particle density of the soil to be 2.65 g cm^{-3}]

(a) 2490 cm^3 (b) 3170 cm^3

(c) 5660 cm^3 (d) 6340 cm^3

86. In the Munsell notation 10YR 6/4, the hue is:

(a) 10 (b) YR

(c) 10 YR (d) 6/4

87. Which of the following statements about soil colour is *true*?

(a) Soil colour has direct effect on plant growth.

(b) Arid soils tend to be grayish to black in colour.

(c) Predominantly soil colours are not pure, but mixtures.

(d) Dry soils are darker in colour than wet or moist soils.

88. As compared to atmospheric air, soil air usually contains several times higher concentration of:

(a) Nitrogen (b) Oxygen

(c) Carbon-di-oxide (d) None of the above

89. Which type of mulching material would result in the warmest soil temperature?

(a) Paddy straw (b) Black plastic film

(c) White (transparent) increases (d) Coir pith

90. The water in the soil that plants can use for their growth is called as:

(a) Gravitational water (b) Field capacity water

(c) Wilting point water (d) Available water

91. Which of the following is plant-available water?

(a) Water filled in micropores only

(b) Water filled in macropores only

(c) Water held at a tension of 3 bars

(d) All the above (a + b + c)

92. The force of attraction between soil and water molecules is:

(a) Adhesion (b) Cohesion

(c) van der Waal's (d) None of the above

93. Which of the following water potentials is *positive,* while the others are negative?

(a) Matric potential (b) Osmotic potential

(c) Gravitational potential (d) None of the above

94. When the water is held in the soil at zero tension,

(a) All the macropores are filled.

(b) All the micropores are filled.

(c) Both the macropores and micropores are filled.

(d) Neither the macropores nor the micropores are filled.

95. At permanent wilting point, plants

(a) Wilt permanently

(b) Wilt temporarily

(c) Wilt temporarily but recover on addition of water

(d) Wilt during day time and recover during night time

96. When a soluble fertilizer incorporated into a soil, it:

(a) Lowers the matric potential of soil water.

(b) Lowers the osmotic potential of soil water.

(c) Lowers the gravitational potential of soil water.

(d) Increases the total potential of soil water.

97. Which of the following forces are responsible for retention of water in soils?

(a) Adhesion

(b) Cohesion

(c) Both Adhesion and Cohesion

(d) Neither Adhesion nor Cohesion

98. The force with which the soil particles retain water was defined in terms of pF scale by:

(a) Schofield (b) Sorensen

(c) Darcy (d) Stokes

99. The water that is held by a soil with a tension ranging from $-^1/_3$ bar to – 15 bars is:

(a) Run-off water (b) Gravitational water

(c) Unavailable water (d) None of the above

100. The pF value of 0 (zero) indicates the following moisture condition in soil:

(a) Oven-dry condition (b) Hygroscopic moisture

(c) Field capacity moisture (d) Saturated condition

101. The downward entry of water into the soil surface is termed as:

(a) Permeability (b) Hydraulic conductivity

(c) Percolation (d) Infiltration

102. Compared to large pores, water in small pores is:

(a) Easier for plant roots to take up

(b) More susceptible to percolate by gravity

(c) Held more tenaciously by the soil

(d) All the above

103. Water flow by capillary movement to plant roots is:

(a) Saturated flow (b) Unsaturated flow

(c) Vapour movement (d) None of the above

104. The movement of water vapour from a zone of high total gas pressure to a zone of low total gas pressure is:

(a) Mass flow (b) Diffusion

(c) Unsaturated flow (d) Saturated flow

105. Darcy's law is related to:

a. Saturated flow b. Unsaturated flow

c. Both saturated flow and unsaturated flow d. Water vapour movement

106. Which of the statements about unsaturated flow of water is *true*?

(a) Water held with potentials greater than -1/3 bar.

(b) Water will move towards the region of higher potential.

(c) Takes place when the soil macropores are filled with water.

(d) Water moves from wetter to drier areas.

107. By definition, soil colloids are:

(a) Inorganic in nature
(b) Organic in nature
(c) Crystalline in nature
(d) Less than 1 or 2 micro-metres in diameter

108. The concentration of a solute on the colloidal surfaces is called:

(a) Adsorption
(b) Desorption
(c) Adhesion
(d) Cohesion

109. A tetrahedral shape is formed when a small atom like silicon is surrounded by four atoms of:

(a) Aluminium
(b) Magnesium
(c) Oxygen
(d) Phosphorus

110. An octahedral shape is formed when an atom like aluminium is surrounded by:

(a) 6 oxygen atoms
(b) 8 oxygen atoms
(c) 6 silicon atoms
(d) 8 silicon atoms

111. The crystal units of 2:1 clay minerals contain:

(a) An octahedral sheet being sandwiched between two tetrahedral sheets

(b) A tetrahedral sheet being sandwiched between two octahedral sheets

(c) An octahedral sheet being sandwiched between two brucite layers

(d) A tetrahedral sheet being sandwiched between two brucite layers

112. A clay in which trivalent cations such as Fe^{3+} and Al^{3+} fill two of the three voids in the half unit cell, the clay is called:

(a) Dioctahedral clay
(b) Trioctahedral clay
(c) Diphormic clay
(d) Triphormic clay

113. Which of the following is an example of isomorphous substitution in clay minerals, leading to the creation of net negative charges?

(a) Substitution of Al^{3+} by Mg^{2+} in an octahedral

(b) Substitution of Si^{4+} by Al^{3+} in a tetrahedral

(c) Both (a) and (b) above

(d) Substitution of OH^- by O^{2-} in a tetrahedral

114. The dominant clay mineral in black soil is:

(a) Kaolinite (b) Illite

(c) Montmorillonite (d) Allophane

115. The dominant clay mineral in red soils is:

(a) Montmorillonite (b) Kaolinite

(c) Illite (d) Allophane

116. Among the following soil colloids, which one has the highest CEC?

(a) Kaolinite (b) Montmorillonite

(c) Illite (d) Humus

117. The amorphous clay mineral is:

(a) Montmorillonite (b) Kaolinite

(c) Allophane (d) Vermiculite

118. Which of the following minerals is non-expanding 2:1 type?

(a) Kaolinite (b) Montmorillonite

(c) Nontronite (d) Illite

119. Which of the following statements about montmorillonite is *false*?

(a) Montmorillonite is an important primary mineral.

(b) Montmorillonite is an expanding type of mineral.

(c) Montmorillonite has a basal spacing of 17 Å.

(d) Montmorillonite has a CEC of 80 - 150 meq/100 g.

120. The particles having the highest specific surface and charge density are:

(a) Organic colloids (b) Clay

(c) Silt (d) Sand

121.The ion with the highest replacing power among all basic cations is:

(a) Na^+ (b) NH_4^+

(c) Ca^{++} (d) None of the above

122.The forces responsible for adsorption reactions are:

(a) van der Waal's (b) Protonation

(c) Hydrogen bonding (d) All the above

123.The cation exchange capacity (in cmol (p^+) kg^{-1}) of montmorillonite clay is:

(a) 3-15 (b) 15-40

(c) 80-150 (d) 150-200

124. The important bases responsible for base saturation are:

(a) Ca and Mg (b) H and Al

(c) Fe and Al oxides (d) Zn and Fe

125. The exchangeable ions (meq/100g) in a soil are: [Ca^{2+}]: 10.0, [Mg^{2+}]: 2.0, [K^+]: 0.26, [Na^+]: 0.09, [NH_4^+]: 1.2, [$H^+ + Al^{3+}$]: 3.6. What is the cation exchange capacity of the soil?

(a) 4.8 (b) 12.4

(c) 13.6 (d) 17.2

126. The exchangeable ions (meq/100g) in a soil are: [Ca^{2+}]: 10.0, [Mg^{2+}]: 2.0, [K^+]: 0.26, [Na^+]: 0.09, [NH_4^+]: 1.2, [$H^+ + Al^{3+}$]: 3.6. What is the base saturation (%) in the soil?

(a) 12 (b) 58

(c) 72 (d) 79

127.The major source of negative charges on humus colloids is:

(a) Carboxylic group (-COOH) (b) Enolic (-OH)

(c) Phenolic group (d) All of these

128.Which of the following soils will have the highest cation exchange capacity, assuming that the amount of organic matter is comparable in all the four soils?

(a) Sandy loam (b) Clay loam

(c) Silty clay (d) Clay

129.The anion exchange capacity increases as the soil pH:

(a) Decreases
(b) Increases
(c) First decreases and then increases
(d) Remains constant

130.The particles having the greatest specific surface and charge density are:

(a) Organic colloids
(b) Clay
(c) Silt
(d) Sand

131.When soil pH is measured by using a pH electrode, it is a measure of which pool of soil acidity?

(a) Active acidity
(b) Reserve acidity
(c) Both active acidity and reserve acidity
(d) Exchange acidity

132.Why is Al^{3+} ion considered to be acidic?

(a) Al^{3+} ions, upon hydrolysis, release H^+ ions.
(b) Al^{3+} ions can exchange with Ca^{2+} and Mg^{2+} ions.
(c) Al^{3+} ions are present only in acidic pH conditions.
(d) All the above.

133. Which of the following factors greatly influence the rate of decomposition of plant litter in soil?

(a) C:N ratio of plant litter
(b) Lignin and cellulose content of plant litter
(c) Soil aeration
(d) Soil moisture

134. What happens when a crop residue with high C:N ratio is added to a soil?

(a) Soil microbial population declines immediately.
(b) The C:N ratio decreases in due course.
(c) There is enough N release to avoid NO_3 depression period.
(d) All the above happen.

135.During the composting process, the C:N ratio of organic materials:

(a) Increases (b) Decreases

(c) Remains constant (d) Fluctuates in cycles

136. Which of the following conditions would favour the accumulation of soil organic matter?

(a) Warm tropical climate

(b) Frequent tillage operations

(c) Good soil aeration and rapid drainage conditions

(d) Fine texture, with high clay content

137. Insoluble fraction of organic matter is:

(a) Humin (b) Humic acid

(c) Fulvic acid (d) Hymatomelanic acid

138.Which of the following statements about soil humus is *correct*?

(a) The CEC of humus colloids increases with increase in soil pH.

(b) One gram of humus will hold less water than one gram of montmorillonite clay.

(c) The humin fraction of humus is soluble in water at any pH value.

(d) Humic substances constitute about 10 to 15% of soil organic matter.

139. The term (Russian) used to denote passage ways and nests of rodents is:

(a) Solum (b) Pedoturbation

(c) Pedon (d) Crotovina

140.Which of the following is not a soil fauna?

(a) Earthworm (b) Nematode

(c) *Eucalyptus* tree root (d) Snail

141.Which of the following is related to soil fungi?

(a) Mycelium (b) Hyphae

(c) Mycorrhizae (d) All the above

142. Which of the following soil organisms account for most of the soil biomass?

(a) Microflora (b) Microfauna

(c) Macrofauna (d) None of the above

143. Which of the following statements is *true*?

(a) Ill-drained soils favour the growth of aerobic bacteria.

(b) "Petrichor", the earthy aroma after a rain shower, is due to N-fixing bacteria.

(c) Actinomycetes are not capable of decomposing chitin and cellulose.

(d) Fungi are more abundant in acidic soil environments.

144. Mycorrhiza is a mutualistic relationship between:

(a) Algae and lichens (b) Fungi and plant roots

(c) Fungi and bacteria (d) Bacteria and algae

145. The symbiotic association of fungi hyphae with roots of higher plants is:

(a) Mycorrhizae (b) Non-symbiotic N fixers

(c) Symbiotic N fixers (d) None of the above

146. The most significant cause for global warming is emission of CO_2 from:

(a) Burning of forests

(b) Oxidation of soil organic matter

(c) Burning of fossil fuels

(d) All the above

147. The only one soil quality indicator that transcends all the three physical, chemical and biological indicator categories is:

(a) Water relations (b) Organic matter

(c) Nutrient cycling (d) Biodiversity

148. Which of the following combinations of soil quality indicators has one each in physical, chemical and biological indicators?

(a) Infiltration, slaking, soil crusts

(b) Soil pH, electrical conductivity, respiration

(c) Aggregate stability, soil nitrate, earthworms

(d) Bulk density, available water capacity, soil enzymes

149.The soil orders so far not discovered in India are:

(a) Andisol (b) Gelisol

(c) Spodosol (d) All the above

150.The majority of the soils found in India belong to the order:

(a) Alfisol (b) Entisol

(c) Inceptisol (d) Vertisol

Key to Multiple Choice Questions

1	c	31	d	61	a	91	a	121	c
2	a	32	c	62	b	92	a	122	d
3	a	33	b	63	d	93	c	123	c
4	b	34	c	64	b	94	c	124	a
5	c	35	d	65	b	95	a	125	d
6	b	36	c	66	b	96	b	126	c
7	a	37	a	67	d	97	c	127	d
8	a	38	d	68	b	98	a	128	d
9	b	39	b	69	a	99	d	129	a
10	b	40	b	70	a	100	d	130	b
11	d	41	a	71	a	101	d	131	a
12	d	42	a	72	b	102	c	132	a
13	d	43	b	73	a	103	b	133	a
14	a	44	a	74	b	104	a	134	b
15	c	45	c	75	c	105	a	135	b
16	c	46	c	76	c	106	d	136	d
17	b	47	c	77	a	107	d	137	a
18	c	48	b	78	c	108	a	138	a
19	d	49	d	79	d	109	c	139	d
20	a	50	b	80	c	110	a	140	c
21	a	51	d	81	d	111	a	141	d
22	b	52	c	82	a	112	a	142	a
23	d	53	d	83	d	113	c	143	d
24	d	54	c	84	a	114	c	144	b
25	a	55	b	85	d	115	b	145	a
26	c	56	a	86	c	116	d	146	c
27	a	57	b	87	c	117	c	147	b
28	b	58	c	88	c	118	d	148	c
29	b	59	d	89	c	119	a	149	d
30	b	60	c	90	d	120	a	150	c

Problems on Soil Density and Porosity

1. Calculate the bulk density of a 400 cm^3 soil sample that weighs 600g and has 12% moisture.

   Oven-dry weight of soil = 600 ÷ 1.12 = 535.71 g

   Bulk density = Mass of oven-dry soil ÷ Volume of soil = 535.71 ÷ 400 = **1.34 g cm^{-3}**

2. Calculate the volume of a soil sample that weighs 600 g and has 10% moisture and bulk density of 1.3 g cm^{-3}.

   Oven-dry weight = 600 ÷ 1.1 = 545.45 g

   Volume of soil = Oven-dry weight ÷ Bulk density = 545.45 ÷ 1.3 = **419.58 cm^3**

3. Calculate the porosity of a soil sample that has a bulk density of 1.40 g cm^{-3}, assuming that the particle density is 2.65 g cm^{-3}. [See § 13.3]

   Porosity % = {1- (BD ÷ PD)} x 100

   = {1- (1.40 ÷ 2.65} x 100 = **47.17%**

4. Calculate the bulk density of a soil sample that has a porosity of 45%. Assume that the particle density is 2.65 g cm^{-3}.

   Porosity ={1- (BD ÷ PD)}

   0.45 cm^3 = {1- (BD ÷ PD)}

   (BD ÷ PD) = 0.55 cm^3

   BD = 0.55 x 2.65 g cm^{-3} = **1.46 g cm^{-3}**

5. Calculate the particle density of a soil sample that has a bulk density of 1.4 g cm^{-3} and porosity of 46%.

   Porosity = {1- (BD ÷ PD)}

   0.46 cm^3 = {1- (BD ÷ PD)}

   (BD ÷ PD) = 0.54 cm^3

   PD= 1.4 ÷ 0.54 = **2.59 g cm^{-3}**

6. Calculate the porosity of a 500 g soil sample that contains 125 g of water when 50% of its pores are full of water.

Oven-dry soil weight = 500 – 125 = 375 g

Volume of soil solids = 375 ÷ 2.65 g cm^{-3} = 141.50 cm^3

Saturated water content = 125 ÷ 0.50 = 250 cm^3 water

Total volume of soil = 141.50+ 250 = 391.50 cm^3

Porosity = {(Volume of air + Volume of water) ÷ Total volume} x 100

= {250 ÷ 391.50}x 100 = **63.86%**

Further Reading

Bockheim, J.G., Gennadiyev, A.N., Hammer R.D. and Tandarich J.P. 2005. *Historical development of key concepts in pedology*. Geoderma, 124: 23 – 36.

Donahue, R. L., Miller, R. W. and Shickluna, J. C. 1990. *Soils. An Introduction to Soils and Plant Growth.* 5th Edition. Prentice Hall of India Pvt. Ltd., New Delhi.

Jenny, H. 1941. *Factors of Soil Formation - A System of Quantitative Pedology.* McGraw-Hill Book Company Inc., NY.

Joffe, J. S. 1949. *The ABC of Soils.* Pedology Publications, Rutgers Univ. Press, New Brunswick, New Jersey.

Soil Science Division Staff. 2017. *Soil Survey Manual.* US Department of Agriculture Handbook No.18.

Soil Survey Staff. 2014. *Keys to Soil Taxonomy*. Twelfth Edition, Natural Resources Conservation Service, US Department of Agriculture.

Weil, R. R. and Brady, N. C. 2017. *The Nature and Properties of Soils.* 15th Edition. Pearson, USA.

Colour Plates

Chapter 3: Soil Forming Rocks

Granite (Coarse-grained)

Siltstone

Gneiss

Diorite (Polished)

Limestone (Chalk)

Marble

Gabbro

Sandstone

Quartzite

Basalt

Shale

Slate

Plate 1: Some important rocks

Chapter 4: Minerals

Olivine

Biotite

Na-Plagioclase

Augite

Muscovite

K-Feldspar

Hornblende

Na-Plagioclase

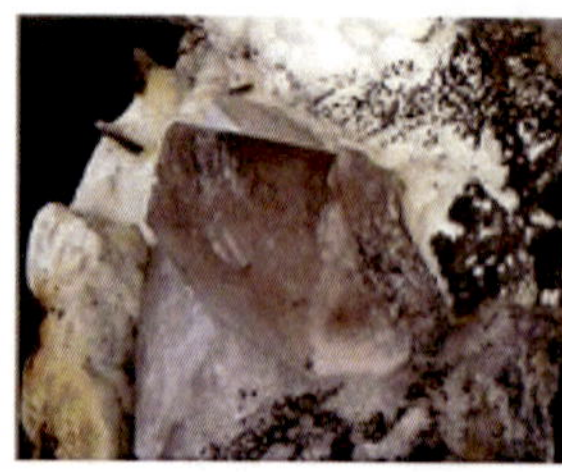

Quartz

Plate 2: Some important soil forming minerals

Chapter 8: Soil Profile

Fig. 8.1: A typical soil profile

A Mollisol profile showing thick dark A horizon with high organic matter content

An Oxisol profile (latosol) in Kerala, with no distinctive horizons and rich in iron oxides

A Vertisol profile with high content of montmorillonite clay

Plate 3: Some important soil profiles

Chapter 10: Soil Texture

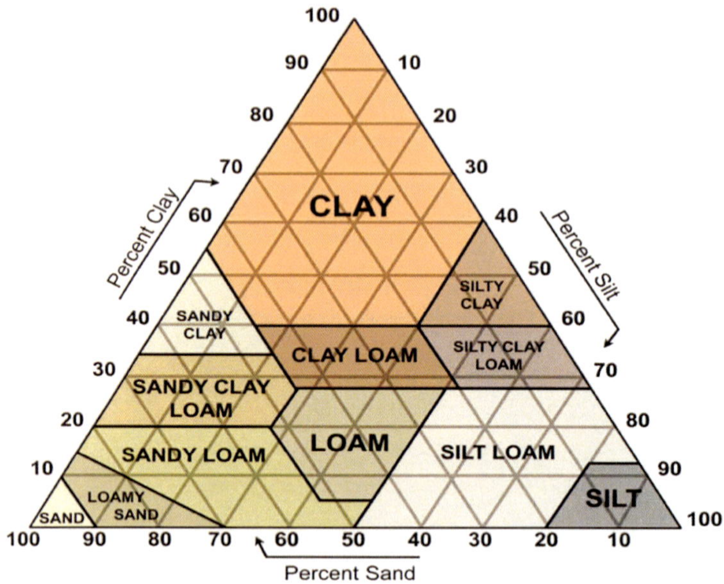

Fig. 10.2: Textural triangle

Chapter 10: Soil Colour

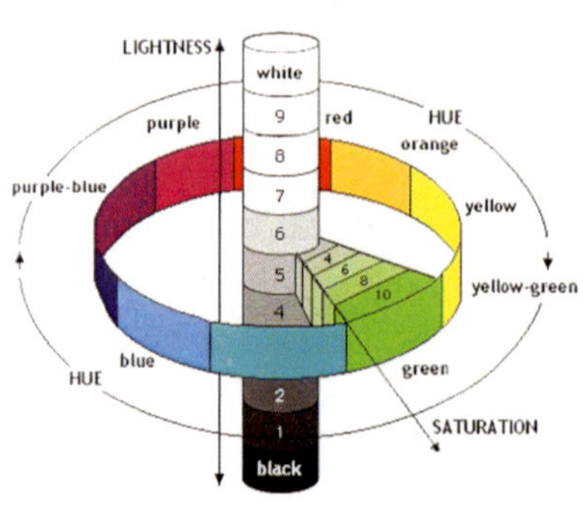

Fig. 14.2: The hue, value and chroma scales of Munsell colour system

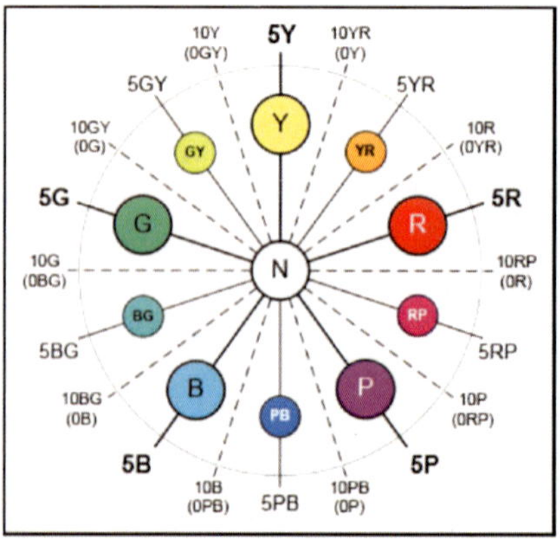

Fig. 14.1: The Munsell hue circle

Index

H

I

L

M

O

P

R

S

T

U

V

W

Z